AF818633

SOTHIS PRESS

Sothis Press aims to completely reimagine the public domain book with an emphasis on universal access to captivatingly designed print editions. We believe that technology will enable every single book ever published to be perpetually accessible in beautiful print form. Sothis is beauty in ubiquity.

www.sothis.press
1 Waterhouse Square, London, ECIN 2ST

Steam and Hot Water Fitters' Text Book

Prepared for the Steam and Hot Water Heating Course at the New York Trade School, With Supplementary Chapters on House Heating, Specifications and Surface Estimating

Thos; E. McNeill

SOTHIS PRESS

ISBN 978-1-0368-0683-5

Steam and Hot Water Fitters' Text Book is in the public domain.

This edition is published by Sothis Press © 2023. The company is registered in England and Wales. The Company Number is 14900389.

Steam and Hot Water Fitters' Text Book

Prepared for the Steam and Hot Water Heating Course at the New York Trade School, with Supplementary Chapters on House Heating, Specifications and Surface Estimating.

BY

THOS E. McNEILL.

ILLUSTRATED

LIBRARY OF CONGRESS
COPYRIGHT.
OCT 22

PUBLISHED BY

DAVID WILLIAMS,

232–238 WILLIAM STREET, NEW YORK.

43403-B2-1

TH6706
M2

Copyrighted, 1896,
by
DAVID WILLIAMS.

4-10706

TO

Charles J. Gillis,

one of the oldest, best known and most highly esteemed members of the Master Steam and Hot Water Fitters' Association of the United States, to whose continued efforts the establishment of the Department of Steam and Hot Water Fitting at the New York Trade School was largely due,

This Book is Dedicated

by

The Author.

CONTENTS.

CHAPTER I.

PAGE.

Tools, Fittings and Pipe, ... 3

CHAPTER II.

General Questions on Heating, ... 16

CHAPTER III.

Low Pressure Steam, ... 21

CHAPTER IV.

Two-Pipe Steam Heating, ... 32

CHAPTER V.

Single Pipe, Low Pressure Steam Heating, . . 45

CHAPTER VI.

Indirect Steam Heating, ... 53

CHAPTER VII.

Hot Water Heating, ... 63

CHAPTER VIII.

High Pressure Steam Heating, ... 73

CHAPTER IX.

High Pressure Steam Heating (*Concluded*), . . 75

CHAPTER X. PAGE.

High and Low Pressure Steam Heating and Power Plant, ... 93

CHAPTER XI.

High and Low Pressure Steam Heating and Power Plant (*Concluded*), ... 106

CHAPTER XII.

Power Fan or Blower System of Steam Heating and Ventilating, ... 110

CHAPTER XIII.

Design, Estimate, Specification. 118

CHAPTER XIV.

Dwelling House Heating, ... 127

CHAPTER XV.

A Successful Heating Job, ... 132

PREFACE.

Early in the summer of 1894 I was invited by the Advisory Committee of the New York Trade School in the Steam and Hot Water Fitting Department to deliver two lectures a week to the class, for twelve weeks of the course, and also to give such aid and advice as might assist in successfully inaugurating the work. Having decided to undertake the work, I discovered that no text book had ever been prepared, and that the course therefore would have to be created. It was with a view to meeting this deficiency that the lectures were prepared, which after delivery were published in *The Metal Worker*, and now appear herewith in book form. It is believed that these lectures will be particularly appreciated by those who wish to master the principles of steam and hot water heating. The definitions in the beginning deal with the appliances, and little by little the scholar is led on until at the close he is told how to figure surfaces and lay out plans and install heating apparatus, with all necessary piping. In a word, the whole subject of steam and hot water heating is covered in a simple way, and it is believed the lectures in the form of a text book will prove of value to all who wish to acquire information in this department. While primarily addressed to young students, the text book will be found of great advantage to those in other lines of trade who wish to take up steam and hot water heating.

THOMAS E. MCNEILL.

New York, June 22, 1896.

CHAPTER I.

TOOLS, FITTINGS AND PIPE.

Question 1. What are the names and applications of the various hand tools used in steam and hot water heating work?

Answer. TOOLS.—The names of the tools used in steam and hot water fitting are: Pipe cutters (Fig. 1) for cutting

Fig. 1.—Pipe Cutter.

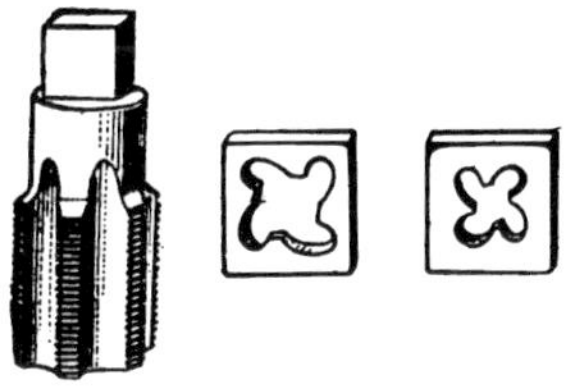

Fig. 3.—Tap and Dies.

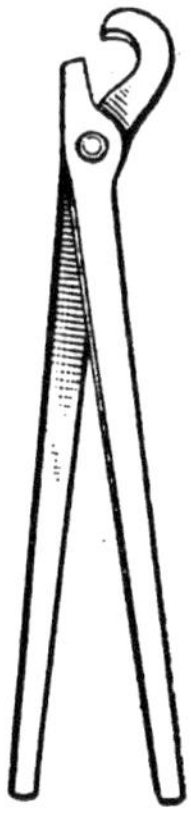

Fig. 4.—Common Pipe Tongs.

Fig. 2.—Stock and Dies.

off pipe; stocks and dies for threading pipes (Figs. 2 and 3), usually from ⅛ inch diameter up to 3 inches or more. The usual thread is standard right hand, but a left hand thread is often used. There are plain or standard tongs (Figs. 4 and 5), a different size being used for each size of pipe. All those of special make are adjustable (Fig. 7) for use with different sizes of pipe. Others for larger sizes of pipe are called chain tongs, or wrenches (Fig. 6). There are many kinds of tongs and wrenches manufactured.

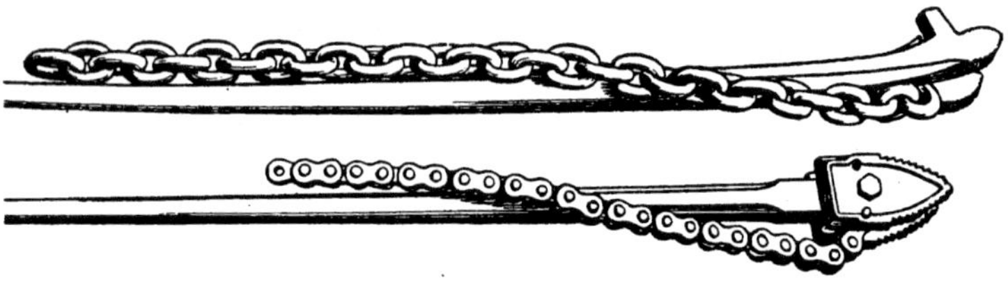

Fig. 6.—Chain Tongs.

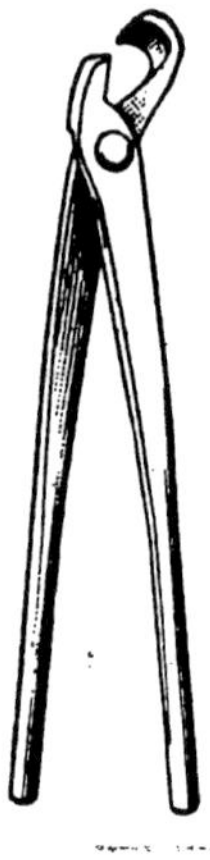

Fig. 5.—Coil Pipe Tongs.

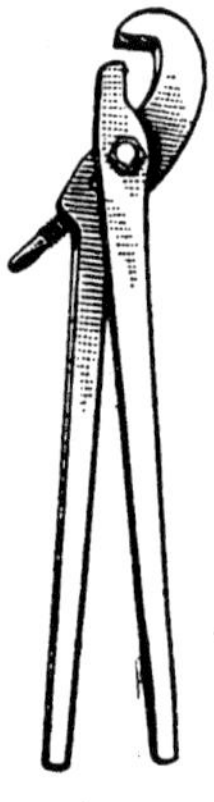

Fig. 7.—Adjustable Pipe Tongs

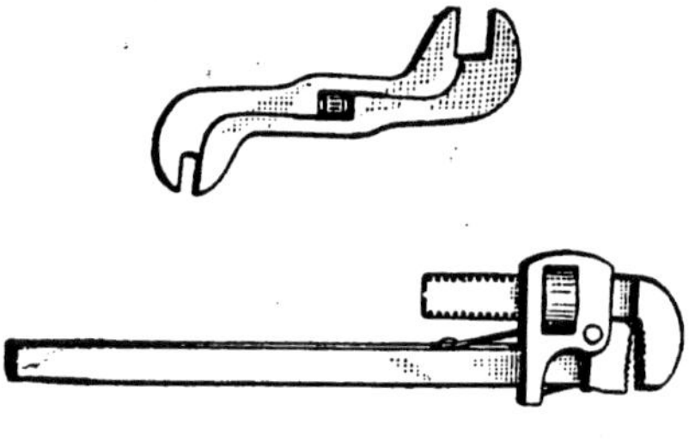

Fig. 8.—Adjustable Wrenches.

Monkey wrenches of various sizes, adjustable wrenches (Fig. 8), hammers, chisels, files, etc.; pipe vises, swivel and angle pipe vises and hinged pipe vises for large sizes (Fig. 9); combination vises, pipe and bench, or machinist work, swivel vise and others, ratchet drill (Fig. 10), etc. (Fig. 11).

Q. 2. What are the names of the principal machine tools used in cutting and threading pipe, tapping holes, etc.?

A. MACHINE TOOLS.—Pipe cutting and threading machines made by different manufacturers from ¼ inch to 2 and up to 12 or 15 inches.

Q. 3. What are the names of the fittings in general use of all kinds?

Fig. 9.—Large Pipe Vise.

A. ELBOWS—The principal fittings are elbows of all sizes (Fig. 14) and reducing elbows (Fig. 15). These are used to change the direction of main pipe or branches, with an easy turn; 45-degree elbows (Fig. 16) of all sizes are used to change direction to angle of 45 degrees, and in combination with other fittings to make easy turns in any direction. There are elbows with a side outlet.

TEES.—Tees (Fig. 17) of all sizes, also reducing tees (Fig. 18), with outlets both on the run (Fig. 19) and the side (Fig. 20); these are used to take out a branch supply and at the same time to continue the direction of main either of the same size or less.

CROSSES.—Crosses (Fig. 21) of various sizes of outlets, also reducing crosses, are sometimes used for continuing the

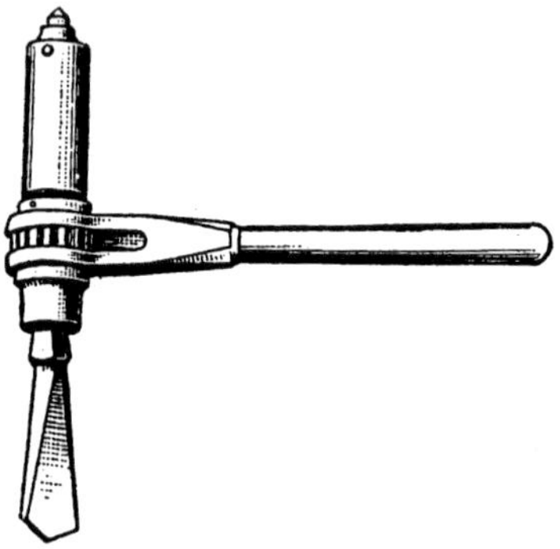

Fig. 10.—*Ratchet with Drill.*

Fig. 11.—*Center Bit.*

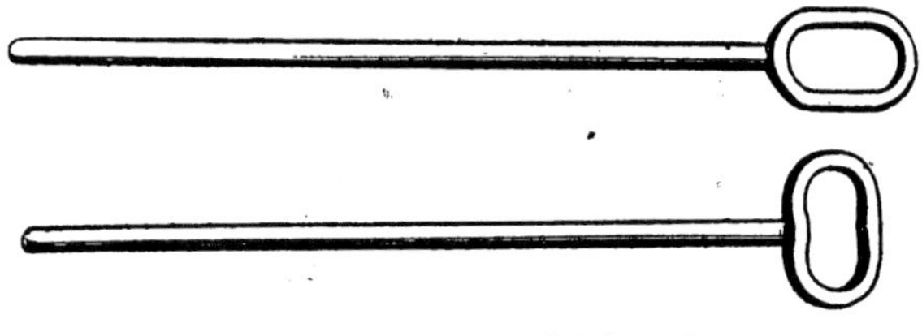

Fig. 12.—*Return Bend Wrenches.*

run of main in the same direction and taking out two supply outlets at the same time.

RETURN BENDS.—Return bends are used of all sizes from ½ inch up to 4 inches; they are made close (Fig. 22) and open (Fig. 23), the first with web between, the latter without.

FLANGE UNIONS.—Flange unions (Fig. 24) are made of all sizes from ½ inch up to 12 inches and are used chiefly

Fig. 13—Pipe Pliers.

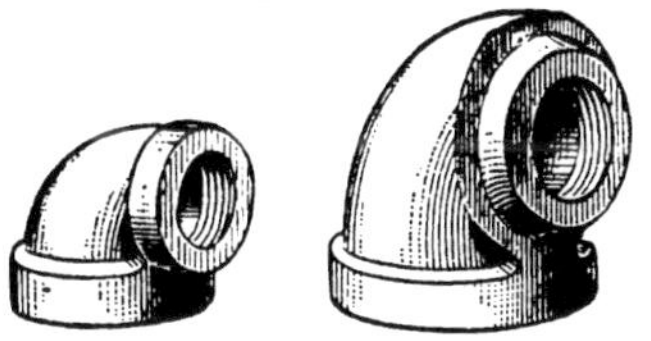

Fig. 15.—Reducing Elbows.

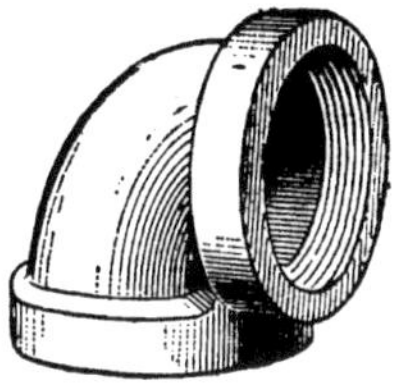

Fig. 14.—Elbow.

Fig. 16.—45-Degree Elbow.

Fig. 17.—Tee.

Fig. 18.—Reducing Tees.

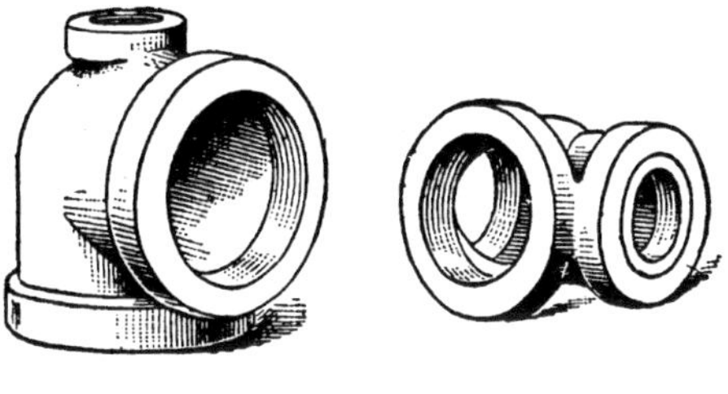

Fig. 19.—Reducing Tees on Run.

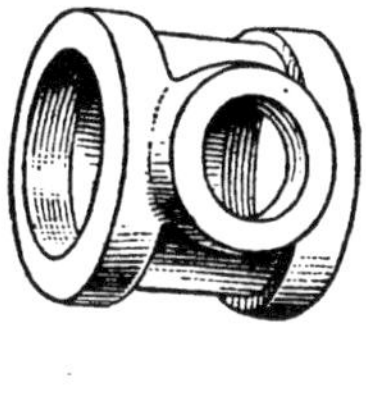

Fig. 20.—Reducing Tee on Outlet.

to connect up large pipes or where the pipes to be connected cannot be sprung.

BUSHINGS.—Bushings (Figs. 25 and 26) are made of all sizes from $\frac{3}{8}$ inch to 12 inches, and are used for reducing the sizes of pipes where fittings for the purpose cannot be obtained, thus making two joints instead of one, as with reducing fittings.

REDUCING COUPLING.—Reducing couplings (Fig. 27), from $\frac{1}{4}$ inch to 12 inches, are used for the same general purpose.

Fig. 21.—Cross.

Fig. 22.—Close Return Bend

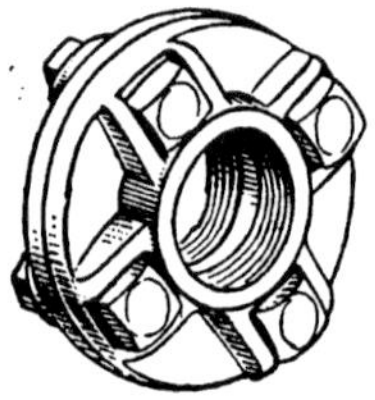

Fig. 24.—Flange Union.

Fig. 23.—Open Return Bend.

Y-BRANCHES.—Y's (Fig. 28) are made of all sizes from $\frac{1}{2}$ inch to 8 inches, and are usefully and advantageously employed, particularly in hot water or steam work, giving an easy change of direction of pipe.

OFFSETS.—Offsets are made of sizes from $\frac{3}{4}$ inch to 6 inches, with offsets 4 inch, 6 inch and 8 inch.

CAPS AND PLUGS.—Caps and plugs (Fig. 29) are made of all sizes from $\frac{1}{4}$ inch to 12 inches, and are used for closing the ends of pipes and outlets of fittings.

LOCK NUTS.—Lock nuts are of all sizes and are generally used for making joints on long threads with some form of packing.

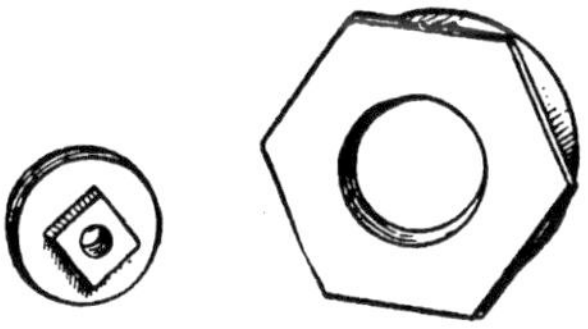

Fig. 25.—Bushings.

Fig. 26.—Reducing Tee with Bushing.

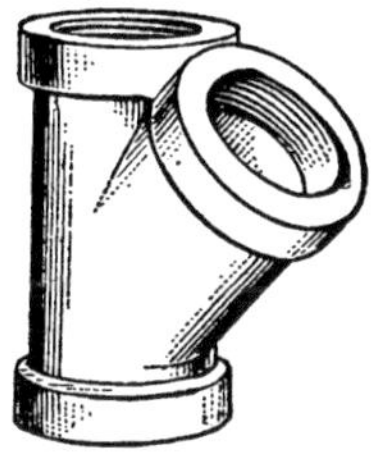

Fig. 28.—Y-Fitting or Y-Tee.

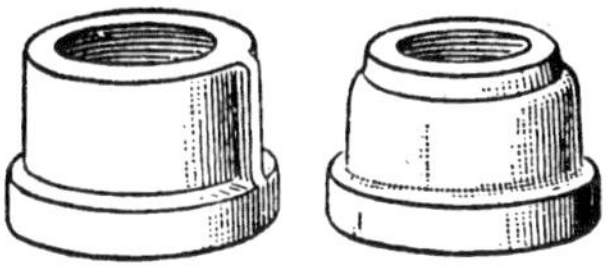

Fig. 27.—Reducing Couplings.

Tapped Plug. Cap. Plug.

Fig. 29.—Cap and Plugs.

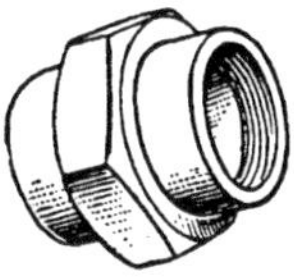

Fig. 30.—Union.

Fig. 31.—Plain Coupling.

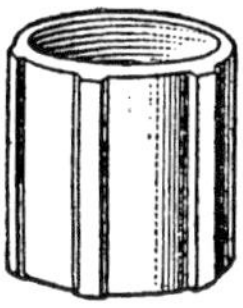

Fig. 32.—Right and Left Coupling.

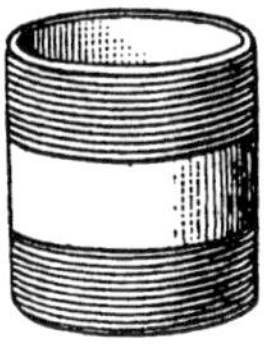

Fig. 33.—Shoulder Nipple.

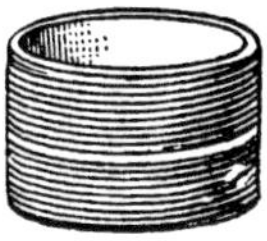

Fig. 34.—Close Nipple.

UNIONS.—Unions (Fig. 30) are made of sizes from ¼ inch to 4 inches, and are used to connect up pipes without spring, and can be easily taken down again.

COUPLINGS.—Couplings are of three kinds, plain (Fig. 31), tapped right hand; ribbed on outside, usually tapped right and left (Fig. 32), and reducing couplings (Fig. 27), mentioned before. The plain are generally used and come on the ends of pipes. Right and left are used to make up joints where ample spring can be obtained.

FLANGES.—Cast iron single flanges are generally used for rests or supports.

NIPPLES.—Nipples are of two kinds, called shoulder (Fig. 33) and close (Fig. 34) nipples, and are threaded right, or right and left hand. The shoulder nipples usually range from 2 inches to 6½ inches long, depending on size. The close from ¾ inch to 4 inches diameter, and from 1½ inches to 5 inches in length.

CEILING AND FLOOR PLATES.—Ceiling plates are made from ¼ inch to 4 inches. Floor plates from ½ inch to 4 inches. The first are used around pipes passing up through ceilings to close and protect the opening around pipes. The last are placed around the pipes where passing through floors, etc., as a finish and protection.

BRANCH TEES.—Branch tees (Fig. 35) are cast iron pipes with tapped outlets on one side, varying in number, sizes and distance apart, with outlets of different sizes on ends, and sometimes outlets on the back or side for special purposes. They are used for making wall circulation coils, with spring pieces of pipe to take up the expansion, and are called, according to arrangement, miter or corner coils.

HOOK PLATES.—Hook plates (Fig. 36), ring plates and expansion plates are all used for supporting the pipes of wall coils, and are usually fastened to wooden strips made fast to the wall.

Q. 4. Are there many special fittings made? If so, what are their names and for what purposes used?

A. ECCENTRIC FITTINGS —Eccentric fittings are not com-

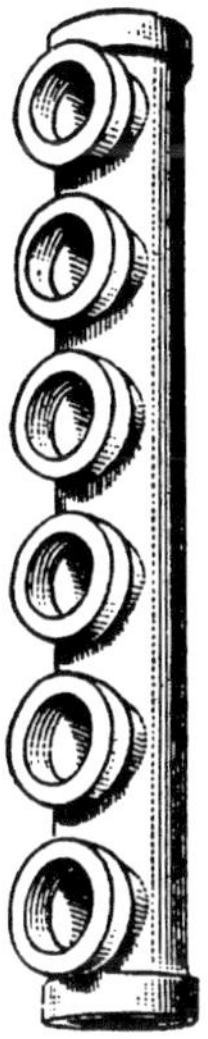

Fig. 35.—*Manifold or Branch Tee.*

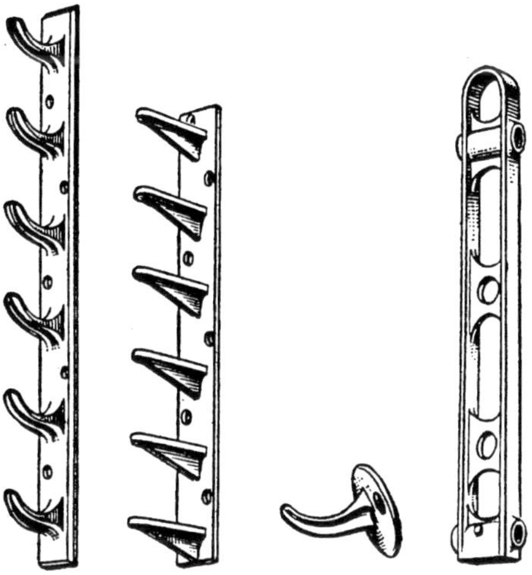

Expansion Plate. Single Hook Plate. Cast Iron Strap.

Fig. 36.—*Hook Plates for Coils.*

Fig. 37.—*Eccentric Coupling.*

Brass Globe Valve Cut Open. Iron Body Globe Stop Valve with Yoke.

Fig. 38.—*Globe Valve.*

mouly used, but are of special value. Eccentric tees and couplings (Fig. 37) may be used in the steam supply mains and other pipes, as they permit of a level bottom in the main pipe and a clear and prompt relief of the water of condensation. Division tees also permit a direct connection of two return water pipes in one line. Double branch elbows are also of use.

Q. 5. Of what material is the pipe generally used in a heating apparatus made? How is it made? Is cast iron pipe ever used for mains or circulations?

A. PIPE.—The pipe formerly used in some classes of hot water heating, and some even now, was made of cast iron; but it was heavy and slow in transmitting the heat, but now in nearly all house work wrought iron pipe is used. Occasionally in factories cast iron pipe is now used for radiating the heat from steam, but the supply mains are usually of wrought iron. In almost all other cases wrought iron pipe is used for mains, risers, returns, etc., but cast iron is largely used for radiators.

Q. 6. What sizes are usually made or kept in stock? What are the various thicknesses?

A. SIZES OF PIPE.—The sizes of wrought iron pipe kept on hand vary from ¼ inch to 12 inches or even 15 inches, and vary in thickness from 0.068 inch up to 0.366 inch—1-inch being 0.134, 2-inch, 0.154, 3-inch, 0.217, 4-inch, 0.237, 5-inch, 0.259, 6-inch, 0.280, 8-inch, 0.322.

Q. 7. Is there any pipe made of extra thickness. and for what purpose used?

A. EXTRA STRONG PIPE.—Extra strong pipe is made of regular sizes from ⅛ inch to 4 inches diameter. It is used for extra heavy pressures, or where liable to rapid deterioration, as connections around boilers for water supply and blow off.

Q. 8. What is the mode of measuring or designating the various sizes of steam pipes? Also of boiler tubes?

A. Pipes and Tube Measure.—The mode of measuring steam and water pipe is by its interior diameter. Boiler tubes are by the outside diameter.

Q. 9. What are the names of the different styles of valves used, and how best utilized? Also of check valves, and how used? Safety and other valves?

A. Valves.—The valves most used on all main steam pipes and around the boiler are either globe (Fig. 38) or angle (Fig. 39) valves; sometimes gate valves (Fig. 40), which

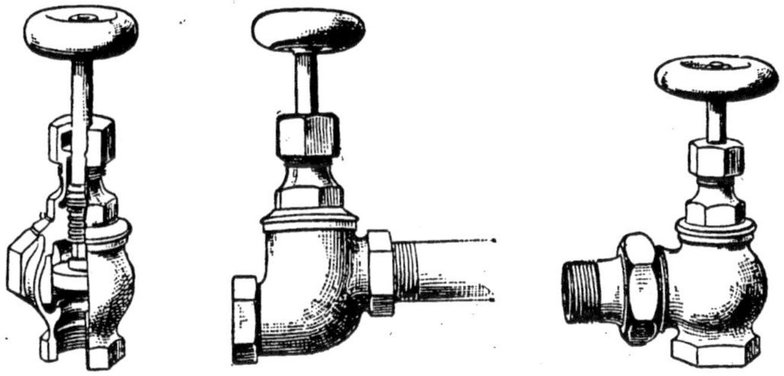

Brass Angle Valve Cut Open.

Brass Opposite Angle Valve.

Angle Radiator Valve with Union Outlet.

Fig. 39.—*Angle Valves.*

are slowly gaining favor; on the bottom of risers, globe valves; around the radiators, chiefly globe and angle or "opposite angle" valves, a special make; gate valves on all hot or cold water pipes.

Check Valves.—Swing check valves (Figs. 41 and 42) are deemed best on the returns to the boiler, pumps, and all pipes conveying water; for all ordinary purposes the common globe checks (Figs. 43 and 44) answer very well.

Q. 10. What are the names of the devices used for hanging or supporting pipes in position, allowing for expansion and contraction?

A. HANGERS.—Pipes are supported by hangers oi various special designs, for ease of adjustment and connection, with freedom of motion for the expansion and contraction of the pipes they support. Various means in the past have been used, such as straps, chain, etc.

Q. 11. What are the different modes used in making up joints when connecting the ends of two or more pipes? What materials are used for special joints?

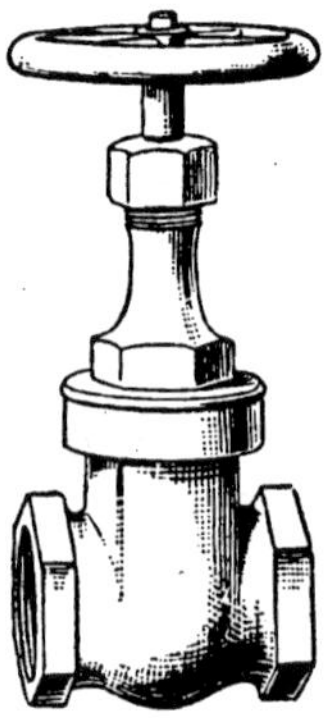

Fig. 40 —*Brass Gate Valve.*

Fig. 41.—*Iron Body Swing Check Valve.*

Fig. 43.—*Common Brass Check Valve.*

Fig. 42.—*Brass Swing Check Valve Cut Open.*

Fig. 44.—*Brass Vertical Check Valve.*

A. JOINTS.—Different means have been used in making up joints. When the threads are good, a little linseed oil may be enough, or plumbago and oil, or red lead and oil. In making up flange unions the joints are made up sometimes with rubber; again with asbestos paper, usudurian, copper, or canvas, oil and red lead. The gaskets are of endless variety.

[Fig. 45 is a shop interior showing fitters' benches made from pipe with pipe vises.]

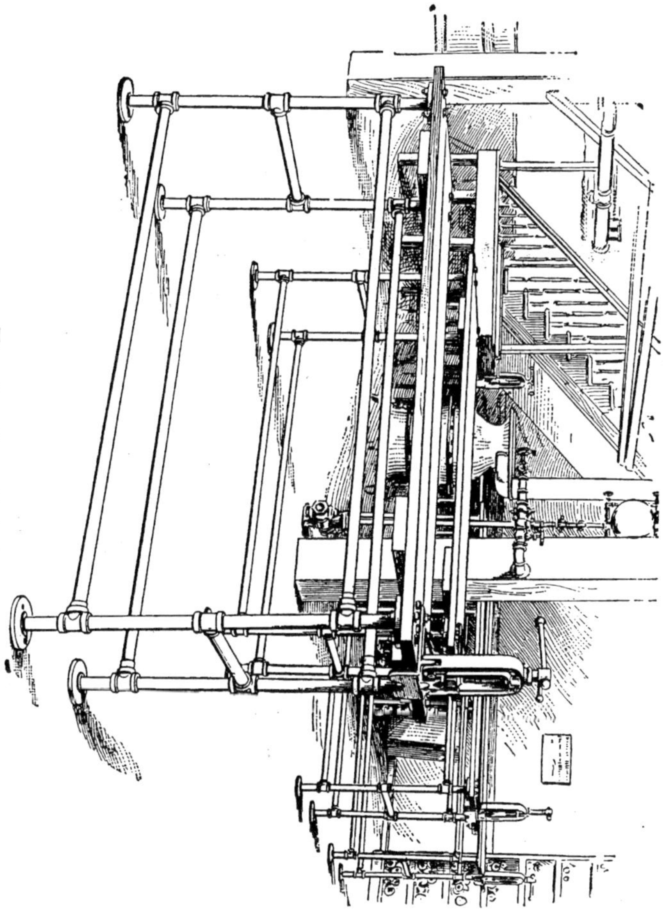

Fig. 45.—Shop Interior.

CHAPTER II.

GENERAL QUESTIONS ON HEATING.

Question 1. What means and devices are generally used for heating dwellings, stores and large buildings of all kinds at the present time?

Answer. Open fires, stoves, hot air furnaces, low pressure and high pressure steam and hot water.

Q 2. Which are considered best, all things duly considered?

A. Low pressure steam or hot water.

Q. 3. How do these compare in first cost and ease of management?

A. The first cost of the hot water apparatus is generally higher; there is little if any difference in the management, as they are both intended to be nearly automatic; both require care and attention in firing.

Q. 4. To what special cases are such best adapted?

A. In variable or moderate climates steam seems best adapted as its circulation and supply can be increased or diminished in less time, and it is more promptly responsive to any sudden requirements. In steadily cold or uniform climates hot water answers all purposes, for, while it needs more time to circulate and thus increase the radiating power, it is more even and maintains an uniform temperature longer.

Q. 5. How are these two systems best applied to their varied uses?

A. (1.) By means of radiators placed in the rooms to be heated; this is called *direct radiation.* (2.) By a combination called *direct-indirect*, where the radiators are placed in the rooms and a fresh supply is brought in from the outer air by means of ducts near to and connected with the base of the radiators. The air is there warmed and rising up passes around the radiator pipes, then enters the room. Also by direct heat from the radiators. (3.) By *indirect radiation*, where the radiators are placed usually in the basement of the house and inclosed in boxes or chambers; air from without is allowed to flow through the heated radiators, then through the ducts, made of tin or galvanized iron, to the various rooms requiring heat, the inflow into the room being regulated by registers and at its entrance into the cold air ducts by dampers. The circulation of the air is caused by its being expanded or rarefied by the heat, and thus made lighter; it then rises up through the fiues. This flow is often quickened by positive or mechanical means, such as fans, etc.

Q. 6. How are these radiators supplied with steam or hot water?

A. By means of a boiler or generator placed, usually, in the cellar, and connected with the radiators by steam supply and water return pipes for the steam apparatus and water circulating main pipes and returns for the water apparatus. In both cases these are connected by branch pipes to each radiator or coil.

Q. 7. What is steam, and how produced for use in steam heating?

A. Steam is the vapor of water, generated in a closed vessel or boiler, and supplied, at any pressure desired, to radiators in the manner previously described.

Q. 8. What is hot water, and how produced for use in hot water heating?

A. Hot water is water charged with heat to or near the boiling point in a closed vessel or boiler, and supplied through circulating pipes to the radiators and coils in the manner previously described.

Q. 9. What are the three terms used in describing the different modes of transmitting heat from one body to the other?

A. Radiation, conduction, convection.

Q. 10. How are these modes utilized in steam and hot water heating?

A. Radiation, as the term is used in heating, is when the heat is transmitted directly to the object having a lower temperature than the source from whence the heat—in this case the radiator—comes and passes through the atmosphere to all objects surrounding it.

Conduction is where the heat passes from one object to another when in actual contact, as from steam to steam pipes, and so on.

Convection is where the heat is conveyed by some medium like air, etc., from one object to another, as in the indirect system of heating.

Q. 11. What is gravity?

A. Gravity is the tendency of all matter to approach the earth, it being the greater mass, as "every particle of matter attracts every other particle in the direct ratio of its mass."

Q. 12. What produces the circulation of steam through the supply pipes and radiators of a steam heating apparatus?

A. Steam being vapor of water, or water expanded by heat, is lighter, therefore moves in all directions and fills all space accessible to it; as the heat it contains passes off, or is conducted off, its volume is reduced and the steam becomes water again, and it being then heavier flows to the lowest point provided for it. In the form of steam it is simply the conveyor of heat by circulation.

Q. 13. What induces the circulation of water in the flow pipes and radiators of a hot water apparatus?

A. As water is heated it expands and, becoming

lighter, rises to the highest point and gives up its heat ; the less expanded or colder descends, becomes heated and again rises, and so on. As the heat passes out from the water by conduction through the iron of the radiator into the atmosphere or objects in a room, the water becomes heavier and descends to the lowest point provided for it. The water simply conveys the heat to the proper point, delivers it and returns for more.

Q. 14. What causes the return of the water back to the lowest point—generally the boiler—in either case?

A. Its greater density, or weight, and the force of gravity in all such systems.

Q. 15. In what direction then should the main supply pipes of a steam heating apparatus incline? Also the return or water pipes?

A. The main supply pipes of a steam heating apparatus should incline downward, from the boiler to the furthest point, or in the direction of the flow of the steam. In the case of branches or side supply pipes, however, this is modified, and they are usually inclined up to the point of delivery. The inclination of the return or water mains is the reverse of the steam, and is downward from the furthest point to the boiler, in the direction of the flow of water.

Q. 16. In what direction should the main supply or flow pipes of a hot water apparatus incline? Also the main return pipes?

A. The main supply or flow pipes of a hot water apparatus should incline upward from the boiler to the point of distribution, and in the same direction for the return pipes.

Q. 17. How should the main supply and return pipes of either system be run as regards their lateral direction?

A. As directly as possible, with due regard to the construction of the building and the appearance of pipes in it, also the location of risers and radiators, avoiding all short turns where possible.

Q. 18. What is expansion? What causes it? How is it provided for in the erection of a heating apparatus?

A. Expansion is the increased separation of the particles of matter, of all kinds, by the action of heat. The reverse, or cold, is the reduction of heat, which is followed by contraction. All materials are subject to like laws in expanding and contracting.

In running pipes containing steam or hot water, due provision is made to meet this in the arrangement of the pipes, by what are called breaks, expansion joints, swinging joints and radial arms.

Q. 19. In designing a heating apparatus for any purpose, what are the principal points to be first obtained, so as to provide for all contingencies and supply the amount of heat required, etc.?

A. About the first information to be obtained is the geographical location of building, of what material constructed, how arranged, what it is to be used for, the points of the compass, the lowest outdoor temperature during winter, the temperature required in the rooms. Where the water supply is to be obtained, if constant and ample. Simplicity in construction, ample proportions of parts for all purposes, without excess, ease of erection, facility for repairs and convenience in management must always be provided for.

CHAPTER III.

LOW PRESSURE STEAM.

Question 1. In planning and proportioning self regulating low pressure steam heating apparatus for dwellings and buildings of moderate size, what parts and appliances are necessary and what are generally used?

Answer. A boiler, either of the horizontal tubular, vertical tubular, sectional, made of wrought iron pipe, sectional of cast iron, or a combination of cast and wrought iron, with all self regulating attachments for controlling air draft, water supply, steam pressure and main controlling valves, etc.

Next, a complete system of main steam supply pipes and branches and return water mains, with all controlling valves necessary.

Next, a set of steam risers and water return pipes, with a valve in each lower branch pipe, except in small houses or when the one-pipe system is used, then one rising pipe answers for both purposes, one valve on the radiator and on each branch pipe below.

Next, where direct or direct-indirect radiation is used, one or more radiators in each room to be heated, with controlling valves, air valves, and pipe connections.

Next, if the indirect is used the radiators are placed in boxes or chambers in the cellar; cold air ducts with dampers and regulators, metal hot air tubes running up to registers in rooms above, must be provided for.

Q. 2. How are the horizontal tubular boilers made, what castings are required?

A. Horizontal tubular boilers are made of rolled plates of steel or iron, built up in section rings and held together by rivets in seams both horizontal and circular; the heads or ends are of flanged steel or iron plates riveted to the shell; in these heads are the holes for the boiler tubes which pass from head to head and are expanded into the head plates by an expanding roller tool; then there is a dome on the top of the shell of like material and make, and all of these heads are thoroughly braced to the sheets. In the dome head, on the shell, or in either boiler head is placed a man hole, and in the lower part of the boiler heads are hand holes, all fitted with covers and fastenings; these are for cleaning out and repairing the boiler. In the upper part of the dome, and on the bottom of back sheet of boiler, are tapped flanges as outlets for steam outflow and return water inlet pipes. On each side of the boiler shell are riveted cast iron lugs for supporting the boiler on brick work. A part of the space above the tubes is for steam.

The castings required for horizontal tubular boilers are sectional cast iron fronts, including smoke box doors, fire and ash pit doors, cleaning out back flue doors, top and bottom, grate bars, bar bearers, arch plate, and flame plate, buck stays, a channel beam for the back arch, and tie rods and anchor bolts, the last three to be of wrought iron.

Q. 3. How are they set, what materials are used, and how laid?

A. Horizontal tubular boilers are set in brick work (Fig. 46). The outer and end walls are made of best hard burned red brick, as a general thing, laid in cement and lime mortar. These walls are carried up above the boiler top, and capped with blue-stone coping. The lining of the furnaces, or fire pots, are of the best fire brick laid on flat in thin fire clay mortar with close joints, and occasionally header courses. The ash pits and a space in front of the boiler are paved with common brick on edge and grouted in with sand and mortar.

There is always one, and sometimes two bridge walls, beneath the boiler; the front one is partly faced with fire brick, and often the space between the two is filled in and paved.

Q. 4. How are the vertical boilers made, what castings are required?

A. The vertical boilers are made of sheets of steel or iron as the others; some are plain cylinders without domes or lugs, set up on end on brick work and above the furnace, others have an annular construction on one end called the fire box formed by an inner sheet riveted to the lower head and the extension of the outer sheet of boiler; the space between these sheets being filled with water.

The castings required for vertical tubular boilers are

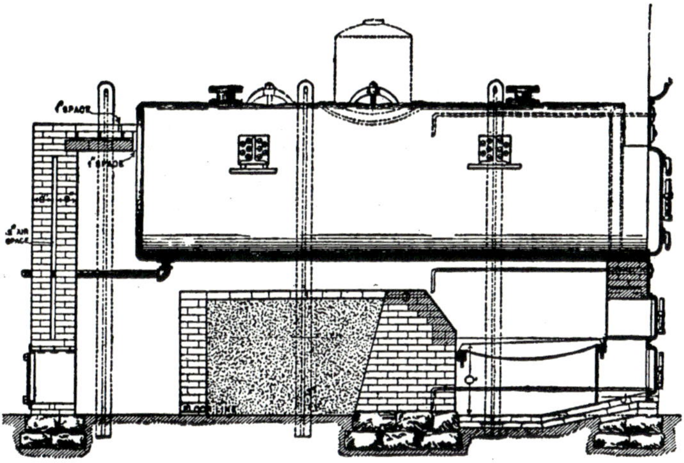

Fig. 46.—Brick Set Horizontal Boiler.

(with or without a fire box) a cast iron bottom and top plate, the latter with a hole in it the diameter of the boiler, fitted with a cover in two parts, a cast iron ring for the grate to rest on, a grate, either dumping or fixed, a fire door and a swinging ash pit door set in frames, sometimes a cast iron base plate for bottom of ash pit and flue cleaning doors at the bottom of the down draft flue.

Q. 5. How are they set, what materials used, and how laid?

A. They are set with the same materials as the horizontal tubular boilers. The simpler class have outer walls of hard brick, and a furnace lined with fire brick, with a space for down draught around the outer shell.

The fire box vertical boilers are set in a similar manner; the outer walls are run up above the top of the boiler, and are capped by a casting, thus forming the smoke box; with this the outer annular flue around the shell of the boiler is connected, the lower end being connected with the smoke flue, thus forming the down draught flue outlet.

Q. 6. How are sectional boilers made, and what castings are required?

A. Sectional boilers are made of almost every possible shape and material, chiefly of a combination of cast parts and headers with wrought iron pipe. One of the earliest styles was in the form of a box coil, so-called; it had cast iron headers with wrought iron coil, made up with cast iron return bends. Others are made of inclined or bent boiler flues expanded into cast iron or steel headers, or wrought iron drums. Others are entirely of cast iron held together by wrought iron bolts, and are of many forms, the object being the same, but the results varied.

The castings required for setting sectional boilers are generally (depending upon their construction) cast iron frames and fronts, smoke, fire and ash pit doors, clearing-out or dust doors for various parts of boilers, arch and flame plates, grate bars and bar bearers, buck stays with tie rods, and sometimes cast or wrought iron beams for carrying the fire bricks covering the top, or cover of fire chamber.

Q. 7. How are they set, what materials are used, and how laid?

A. The settings, in most cases, are only an inclosure of the various parts by outer walls of hard common brick, and furnace and side lining of fire brick, covered on top by fire brick, supported by iron beams.

Others of these boilers are what are called portables—that is, without setting and easily moved.

Q. 8. What are the parts of the self regulating feed apparatus, how made; how connected up with boiler and water supply?

A. The parts of a self acting feed water apparatus (Fig. 47) are an oblong cast iron receptacle or shell, sometimes round, or flattened on the sides and irregular in shape. A lever of brass with an air tight hollow copper sphere is fastened on one end, and two pivoting pieces at the other end; beyond these is a projection with a recess to hold a rubber disk, or valve; on the outer side of this cast iron shell is sometimes attached a glass water gauge; on the top is a hole tapped for the steam connection; on the bottom is another tapped for the feed to the boiler; at the other end, below, is another tapped hole for tne cold water supply pipe.

This is set up by the side of the boiler, sometimes resting on a flange on the floor, or supported by proper pipe connections, etc., the centre of the apparatus being on a line with what is desired to be the water line of the boiler. Pipe connections are then made between the top of the casting and the steam space of boiler, the bottom with the lower part of the boiler direct, or with the main water return pipe near the boiler, and inside of all controlling valves,or checks; the opposite end of casting is connected with the water supply pipe, a cross connection is made between this supply pipe and the discharge pipe to the boiler, with proper controlling cocks or valves in each so that a direct feed to the boiler, if necessary, may be made, or through the self regulating valve, as may be desired; each part should be under complete control, by means of cocks or valves, so that repairs may be made without stopping the apparatus.

The water being admitted by the supply pipe passes through the regulating valve into the casting, from thence through the discharge pipe into the boiler. As soon as the water in the boiler rises to a certain level, which is the same as in the feeder, then the copper ball rising with it actuates the lever, closes the regulating valve and shuts off the water supply. As soon as the water falls below this

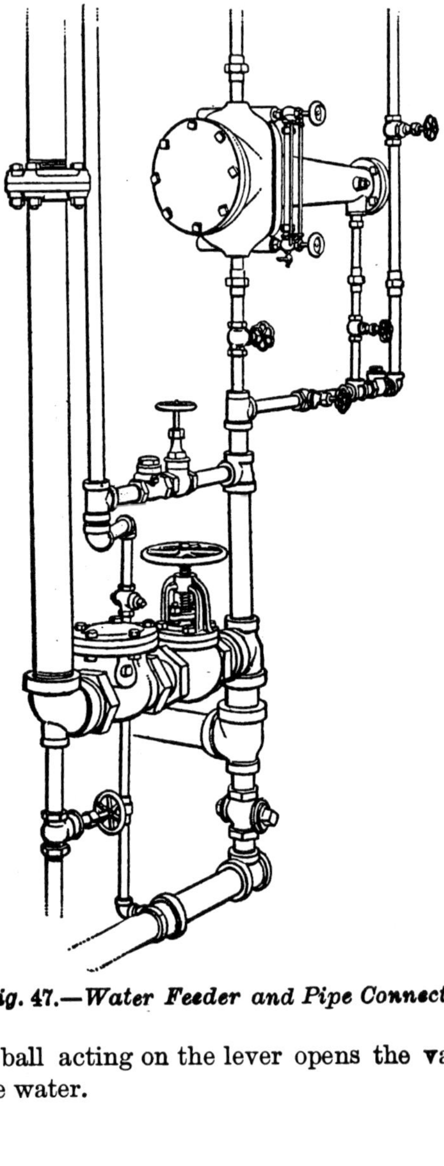

Fig. 47.—Water Feeder and Pipe Connections.

point the ball acting on the lever opens the valve and admits more water.

Q. 9. What are the parts of a draught regulator? How made? How connected up with the boiler and damper?

A. The parts of a draught regulating damper are parts of two hemispherical castings with a flange on each, and a forked projection on one; a small thin casting with a depression in the centre, a pin with a fork at one end, a lever with three small holes through it, one at each end and another near the middle, two small pins, a counter balance weight, and a cup shaped diaphragm of rubber.

The small casting with depression on it is fastened to the centre of rubber diaphragm; this is placed between the two hemispheres, and then they are bolted together, the forked pin is put through a hole in the upper hemisphere and rests in the socket or depression. In casting on the rubber diaphragm, the lever is put in the forks in the top of casting, the pin passed through it, a chain attached at one end of the lever and the counter weight on the other.

The bottom of the lower casting is connected with the steam space of the boiler by means of a siphon pipe, and rests generally on the top of the boiler, or near it; the chain on the end of lever is connected with the lever of the balanced or cold air check damper in the smoke pipe and the swinging ash pit door.

Q. 10. How are the parts of the glass water gauge made? How are they connected up and used?

A. The parts of a glass water gauge are generally a cast iron stand pipe, or water column, with holes tapped top and bottom, three holes for gauge cocks tapped on one side, and two holes on another side, one near the top and the other near the bottom, tapped for the brass trimmings and pipe connections.

The gauge proper is made of brass, with a receptacle for the glass, which may be of varied length, with valves on the top and bottom.

These parts are screwed into the stand pipe, the glass put in place, and made steam tight by rubber washers; three regular gauge cocks are screwed into their places, a pipe connection is then made between the steam space of the boiler and the top of water column and the water space of boiler and the bottom of water column, with a controlling valve in each. The water line of boiler coincides with a line about the centre of water column: sometimes it is varied.

Q. 11. What are the uses of a steam gauge, how is it made, and how connected with the boiler?

A. The use of a steam gauge is to show the pressure of steam which is being carried in the boiler.

The parts of a modern standard steam pressure gauge are a cast iron or brass case, a flattened metal tube curved to a certain shape, and called the "Bourdon tube spring" (from the name of its inventor), several levers, and a pointer. The tendency of the pressure of steam when admitted into this pipe is to straighten it. This is the actuating force, and the movement produced is transmitted by the levers, etc., to the pointer; as the pressure varies, the levers are moved; the amount of motion in the spring and the pointer showing on the graduated face of the gauge front. Sometimes a double spring is used for special purposes, also a corrugated steel diaphragm is used, which is placed between two disks or hemispheres, and the motion transmitted directly to the pointer. The gauge is connected with the steam space of the boiler by means of a pipe and a siphon.

Q. 12. What are the uses of a safety valve? How is it made? How should it be placed and connected up with the boiler?

A. The use of the safety valve (Figs. 48. 49 and 50) is to relieve the steam pressure in the boiler when it passes beyond the point desired by blowing it out into the atmosphere. For low pressure boilers it is usually like an ordinary angle valve with the spindle running out of the top, without packing (Fig. 48); on this rests a ball or disks of any weight required for the steam pressure it is desired to carry. Others have a lever and weight (Fig. 49). It should be placed near the top of the boiler and connected directly with the steam space without a cock or valve in the connecting pipe.

Q. 13. What valves are required in the main supply

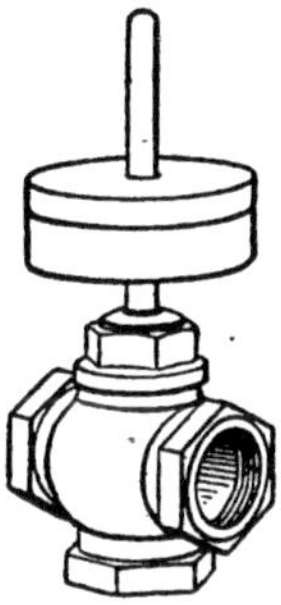

Fig. 48. — *Brass, Dead Weight Low Pressure Safety Valve.*

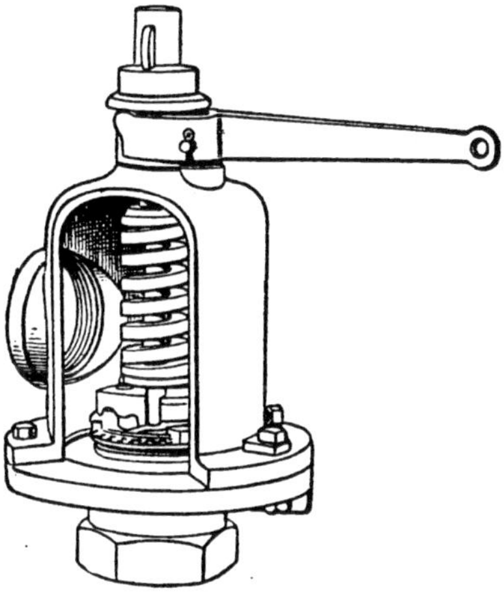

Fig. 50.—*Section of Pop Safety Valve.*

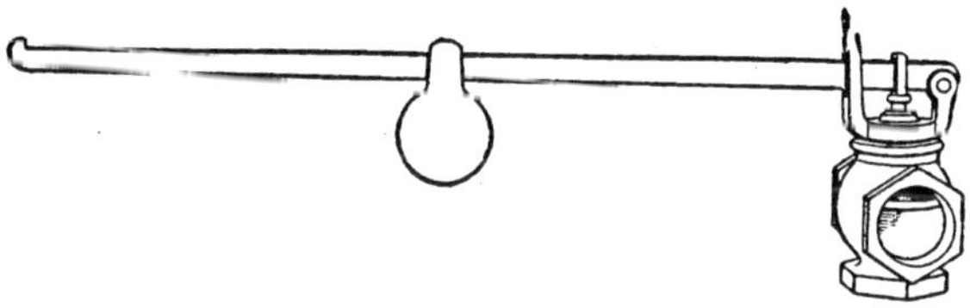

Fig. 49 —*Brass, Lever High Pressure Safety Valve.*

pipes for all returns to the boiler, blow off pipe, water feeder, glass water gauge, or any other parts, such as pass-bys, etc. ? How arranged in pipe ?

A. When using very low pressures it was the practice of many to place no controlling valves on either the steam or return mains, but the general practice now in buildings of moderate size is to place a globe or angle valve (Figs. 38 and 39) in the main steam pipe and globe or gate valve (Fig. 40) and a swinging check valve (Figs. 41 and 42) in the return, near the boiler; on the blow off either a plain brass cock (Fig. 51) or an asbestos packed one (Fig. 52) is now used; in the water feeder connections about five steam cocks or valves and one check valve (Fig. 43) are used; on the water column a valve top and bottom, and a blow out valve or cock at the lowest point; on the pass-by from the return main a blow out cock or valve.

Fig. 51.—Brass Steam Cock.

Fig. 52.—Section Asbestos Packed Steam Cock.

Q. 14. Should each part be under controlling cocks or valves and independent of the boiler?

A. Yes; in every way, so that repairs may be made without interference with the working of the apparatus, the safety valve and the steam gauge always excepted.

Q. 15. Are the parts above mentioned and described necessary and applicable to all forms of low pressure steam heating boilers and apparatus?

A. The parts herein described are considered essential for the best work.

Q. 16. How should the parts of pipes passing through brick work or in any way exposed to extra heat be protected?

A. The parts of pipes passing through brick work or

hot air flues, when containing either steam or water, such as the feed water pipes, the blow off pipes, and sometimes parts of steam mains, should be protected by sleeves made of larger size iron pipe, or by some non-conducting material, or both, depending upon the temperature of the hot air or gas.

CHAPTER IV.

TWO-PIPE STEAM HEATING.

Question 17. What appliances are used for conveying the steam generated in the boiler to the radiators where it is utilized?

Answer. The appliances used for conveying the steam from the boilers to the radiators are the steam supply and return water of condensation pipes, such as horizontal mains and their branches to the rising steam and return water lines, the branch steam supply pipes to the radiators and their water returns, all to be run with great care and as straight as possible without deflections, known as traps and pockets. (Figs. 53, 54 and 55 show basement, first and second story plans of a residence heated by the two-pipe steam direct radiation system.)

Q. 18. How are the steam and return mains run, what means are used for relieving the steam main of water of condensation, and what inclination is given to the pipes?

A. The steam mains are run from the highest point near the boiler to the point nearest the risers or to the radiators on the first floor, with a slight inclination from the boiler to the furthest point to be supplied. The rate of inclination is usually about ¼ inch in 10 feet, sometimes more and again less, depending on circumstances. Where the run is very long it sometimes becomes necessary to make a second rise to the highest point practicable and again commence another run with proper inclination, etc., as before. This rise or break also provides a means of taking up the expansion in the pipe. The return mains are generally run parallel with and near the steam mains, but with ample inclination toward the boiler, the reverse of the steam (Fig. 56). Outlets are left at suitable points in both mains, as required, for the branch pipes to risers or radia-

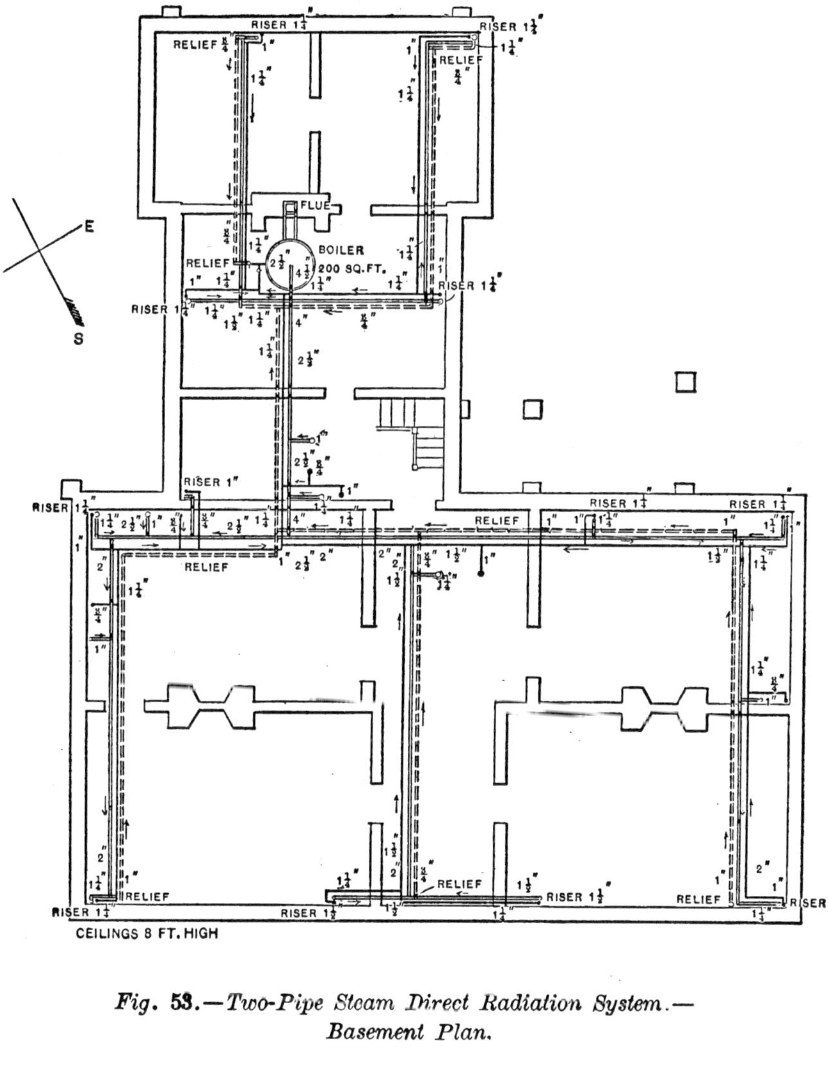

*Fig. 53.—Two-Pipe Steam Direct Radiation System.—
Basement Plan.*

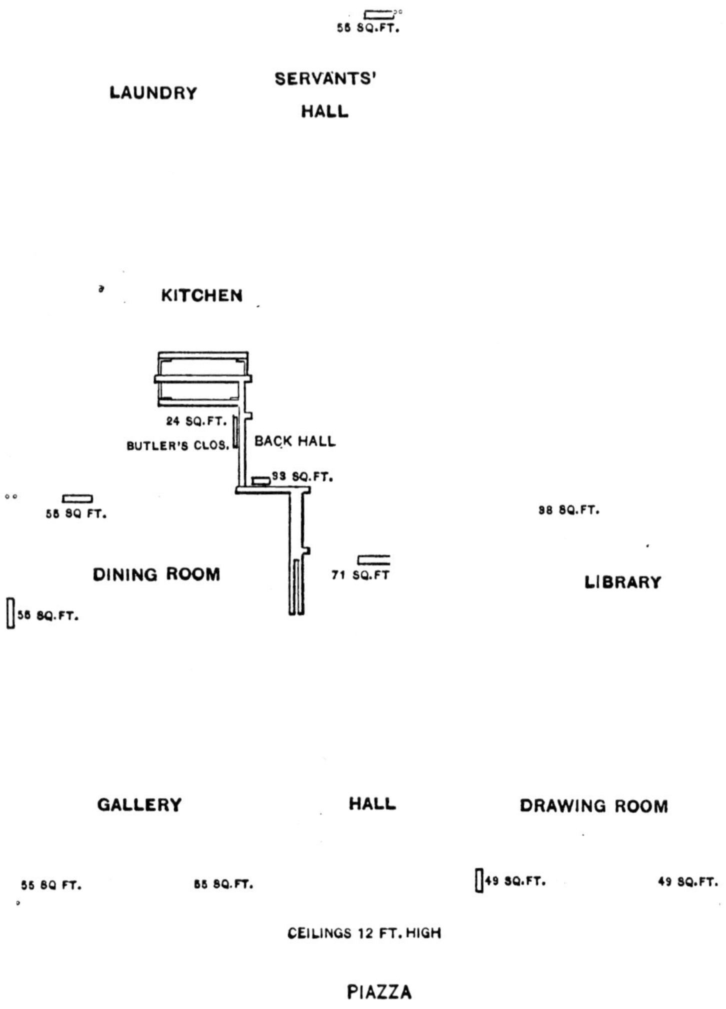

Fig. 54.—Two-Pipe Steam Direct Radiation System.—
First-Floor Plan.

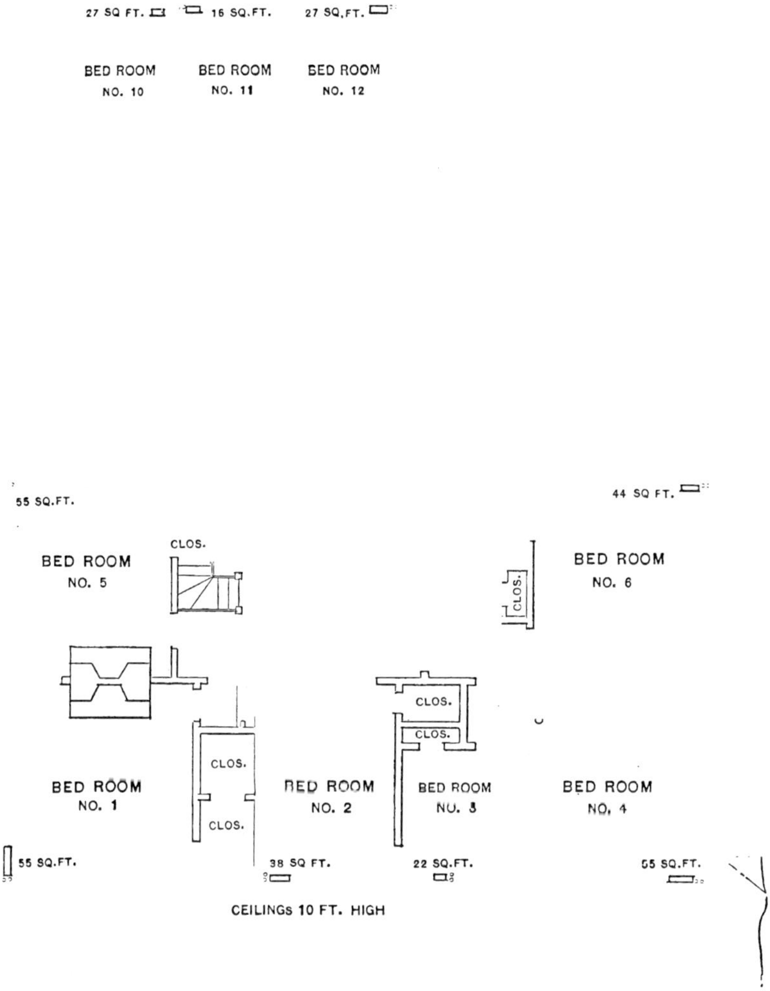

Fig. 55.—Two-Pipe Steam Direct Radiation System.—Second-Floor Plan.

tors and their returns. A relief pipe is taken out at the end of the steam main, or wherever a break or rise is made or a change of inclination becomes necessary. The relief pipe in low pressure steam work is sometimes taken at once into the return water main, when that is run below the water line of boiler (Fig. 57) or by a separate pipe with its valves and checks when taken from the end of the steam main;

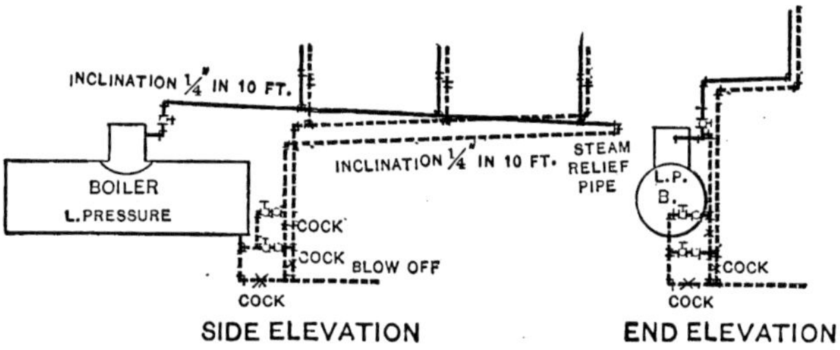

Fig. 56 —Two-Pipe System with Overhead Return.

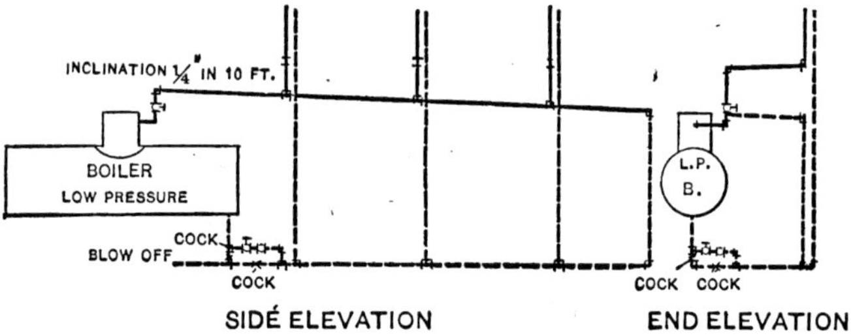

Fig. 57.—Two-Pipe System with Submerged Return

this is run parallel with it and back to the boiler with a separate connection to the boiler, or into the return main inside of the main check and globe valve.

Q. 19. How are the risers run, supported and protected? How many are used? Are valves placed below the risers in the branch connections?

A. The risers are run from the cellar or basement up to the top floor or room to be heated. They are sometimes run in channels made in the walls to receive them, and these are covered over with iron lathing, or wire cloth, plastered over, boarded up or protected by fire proofing, etc. Sometimes they are run in the room clear of the main house walls and are exposed or inclosed with fire proof material or boxed in with sheet metal. The plan most used now is to run them near the walls and leave them exposed and properly finished. This is the easiest and in many respects the best way, as they can be readily reached for repairs and any defects which may develop can be detected. Those run in recesses or flues in exposed walls are usually well felted and protected, the fastening being annealed copper wire.

The risers are supported in their positions by clamps fastened in the wall; short risers are supported from below; in buildings of moderate hight in the middle; in very tall buildings by breaks in the pipe, sometimes one or more, to provide for expansion and contraction, and at the same time a support is obtained between the floors.

Two risers are generally used, one for the steam supply the other for the return water, except in what are called one-pipe systems, where one pipe answers both purposes (a special chapter will be devoted to this system in the future).

Valves are placed below the risers in the branch pipes near to the mains and of the same size as the risers, also on single or one-pipe systems, so that the risers may be shut off for repairs while steam is on the house, but in small two story houses using low pressure steam they are usually omitted.

Risers should be run as near the radiators as the spring of pipes for expansion will admit; long branches are objectionable; it is better practice to put in more risers.

Q. 20. What are the supply and return pipes from the risers to the radiators called? How are these branch pipes run and how relieved of water?

A. The supply and return pipes from the risers to the radiators are called branch pipes, and should be run with great care so as to secure ample drip to and from the radiator and the riser, and make the best arrangement to secure freedom of expansion for both sets of pipes, at the same time to produce a neat and unobjectionable arrangement. Sometimes the branch pipes are run under the floor with connections coming up through it. In wooden buildings and those with wooden floors this can be done, but in fire proof buildings it is very difficult. Sometimes they are run above the floor either back or front of or beneath the radiators, when high legs are used. There are many ways and it requires good judgment and skill to select the best for each case. The inclination of the short steam branch supply pipes from the riser to the radiator or coil valve is sometimes downward toward the radiator. If, however, more than about 6 feet in length, in most cases it is upward to the valve, but downward to the radiator to allow the water condensed in the pipe to drain both ways ; the latter is the best practice. The return branch pipe should always incline downward to the riser.

Q. 21. How are radiators made, and of what material?

A. Radiators are made of wrought iron pipe or cast iron pipe in sections or loops arranged vertically in rows. The principles of construction and circulation of steam are the same in all forms, the difference being in the material used and the shape of parts, each maker claiming some advantage.

The essential parts are a base, or its equivalent, into which pipes are screwed. The pipes in the first form used were of wrought iron screwed into the base and connected at the top by a return bend. Then came what is known as the Nason radiator—it being the invention of Joseph Nason of New York City. This is a single pipe, divided by a sheet iron diapraghm running from the bottom of the pipe up to within a short distance of the top; this arrangement made it practically two pipes in one. Then came the Bundy cast iron pipe. This is a double pipe, united both top and bottom and screwed into the base.

Another form which covers the same ground is a cast iron section, made up of two or more connected barrels or pipes, cast in one piece as a unit, and these united with screwed nipples at the bottom for steam, and both top and bottom for hot water. These form the radiator; the pipe connection at the bottom being large, forms a substitute for the regular base.

Q. 22. What are the different kinds of radiators which are now generally used?

A. The radiators commonly used at the present time are chiefly of cast iron, and made up of sections. There are many manufacturers, each maker having a number of different forms and styles, but the principle of action is about the same. The condensation of the steam on the inner surface of the pipe causes an upward flow of steam from the base or other source of supply; this causes the so-called circulation in the radiator. The air which may be in the radiator when the steam is first turned on is forced forward and goes to the last pipe or section in the radiator, where the air valve is connected.

Q. 23. What is the difference between what is generally known as a radiator and what is known as a wall or other coil?

A. The difference between what is generally known as a radiator and what is known as a wall or other coil is, the radiator is a vertical arrangement of pipes as above described, the coil is horizontal.

Q. 24. Of what material are coils made, and how constructed and arranged and best utilized; also their general advantages?

A. Coils are made usually of wrought iron pipe of from 1 to 2 inches in diameter, and are called return bend coils (Fig. 58), miter wall coils (Fig. 59), corner wall coils (Fig. 60), and box coils (Fig. 61). They are all among the first forms of radiating surface used for plain work generally, although the return bend coil is sometimes used in the better class of work now. The return bend coil, as its name indicates, is made up of uniform lengths of wrought iron pipe connected together alternately on the ends by return bends, thus forming a continuous pipe from top to bottom. The

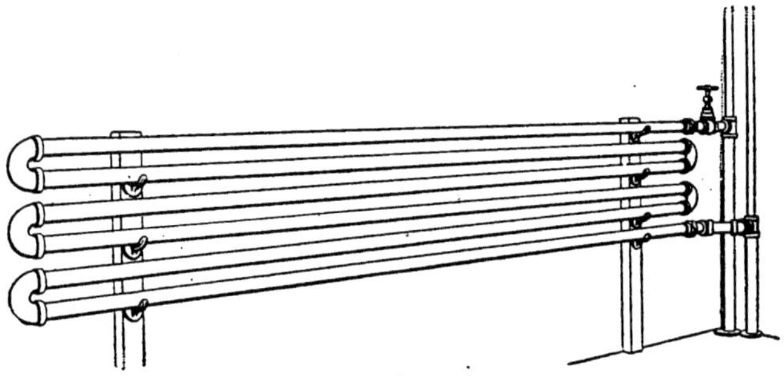

Fig. 58.—Return Bend Hot Water Coil.

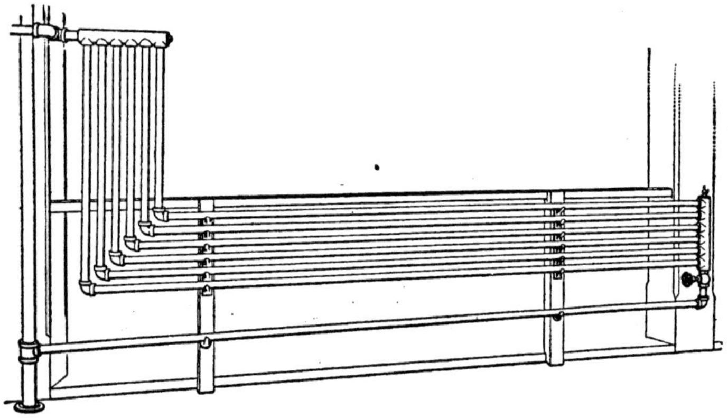

Fig. 59.—Miter Coil.

steam supply is connected to the top pipe and the water discharge to the bottom pipe. The coil generally rests on a series of hooks or racks or is confined by iron straps made fast to wooden battens usually nailed to a wall. They may be of single, double or treble rows.

Box coils (Fig. 61) are made up of a series of the above

described coils united at the top and bottom by a branch tee or header, to which the steam supply and water return pipes are connected. The series of coils are held together

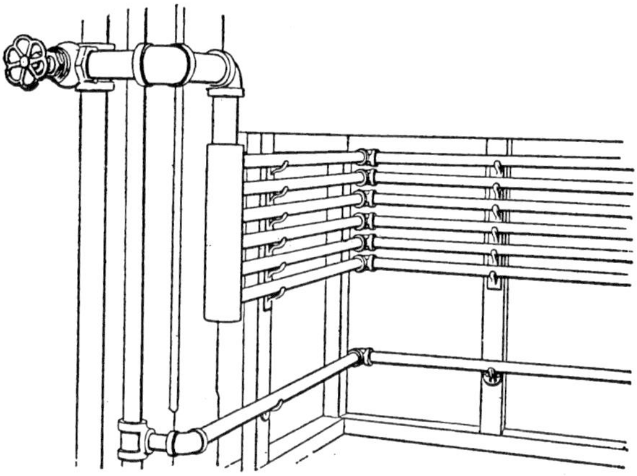

Fig. 60.—Corner Wall Coil.

Fig. 61.—Common Box Coil.

by iron straps with bolts and nuts, the whole resting on feet on the floor.

Wall coils are made of wrought iron pipe with a branch tee or header on each end. In order to provide for expansion, near one end of the coil are elbows and short pieces with a change of direction of run to right angles; these short pieces of pipe are called spring pieces, and by their elasticity in giving a little they take up the expansion of the long pieces of pipe and prevent warping and breaking. When near a corner in a wall these short pieces make the break at the corner. When otherwise located they are turned up or down on the wall, thus producing the same effect.

Such coils as above described are very effective heating surface, and being generally painted black they are good radiators of heat. The horizontal position of the pipe and their ample separation give the rising currents of air a chance to reach the pipe surface. They are generally placed in stores, factories and buildings used for working purposes.

Q. 25. Where should the heating surface in a room be placed to produce the best effect? Which is the coldest side of a room generally?

A. The heating surface in a room should be placed near the coldest side, which is usually the north and west, or near a window, or wherever cold is likely to enter, so as to heat the air up quickly and thus prevent cold drafts.

Q. 26. What sometimes prevents the heating up of a radiator? How is the air expelled?

A. Air will sometimes prevent the heating up of a coil, as it will leak through the joints into the radiator when it is cold, and a partial vacuum is formed by the condensation of the steam remaining in it after the main supply is shut off. When live steam enters again it pushes this cold air forward and compresses it into some pipe or corner, and this prevents the steam from entering or doing its full duty. Generally it collects at a point opposite to where the steam enters or near the return pipe. An air cock or valve, operated either by hand or automatically, is placed at this point and the air is allowed to escape.

Q. 27. How are automatic air valves made? How do they operate? How are they connected up?

A. Automatic air valves were originally made of iron rods and brass tubes so arranged that when cold the valve was slightly open; when steam was turned into the radiator it forced the cold air ahead of it out of the open valve, but the moment the hot steam touched the sensitive brass it expanded and closed the opening and remained closed until air should again collect in it and become cold from want of circulation; then the metal would contract, the valve open and the air escape. Many modifications have been made recently, the principle of action, however, being the same, expansion and contraction by hot steam and cooler air.

Q. 28. How is the steam supply to and the water of condensation return from the radiator controlled?

A. The steam supply to and water of condensation return from the radiator are controlled by radiator valves. The globe corner, opposite angle, angle or gate pattern are used.

Q. 29. What kind of radiators are usually used for direct-indirect heating, and where are they located in a room?

A. Vertical wrought iron or cast iron pipe radiators are used for direct-indirect heating. The space below the usual radiator base is inclosed and the base has holes through it between the pipes. The cold air from the outside of house is brought in through an opening, or duct, of galvanized iron or tin, with a damper in it to regulate the air supply to radiator. The radiator is usually located near a window, as being best for heating purposes, as well as for arranging the cold air ducts.

Q. 30. Is there any means of obtaining a fresh air supply to a room when the regular direct radiator is used?

A. There is a mode of obtaining fresh air for a room when direct radiation is used. The radiator is placed in front of a window, the lower part of the sash is either perforated with holes, which can be closed by a sliding damper, or a damper made for the purpose is let into the frame. The fresh air is thus admitted and the heated air from the radiator rises up and mingles with and tempers it. Another mode is sometimes used: A strip of wood about ½ inch thick by 4 inches wide, and in length the full width of the window, is placed in front of the lower part of the sash, leaving a narrow space between the sash and the strip of wood. When the sash is slightly raised the air enters below it and is deflected upward by the strip of wood. The heated air rising up from the radiator mingles with and warms it to the desired temperature. By raising or lowering the sash, more or less cold air may be admitted.

CHAPTER V.

SINGLE PIPE LOW PRESSURE STEAM HEATING.

Question 1. What is a single or one-pipe system of steam heating (see Figs. 62, 63 and 64, showing basement, first and

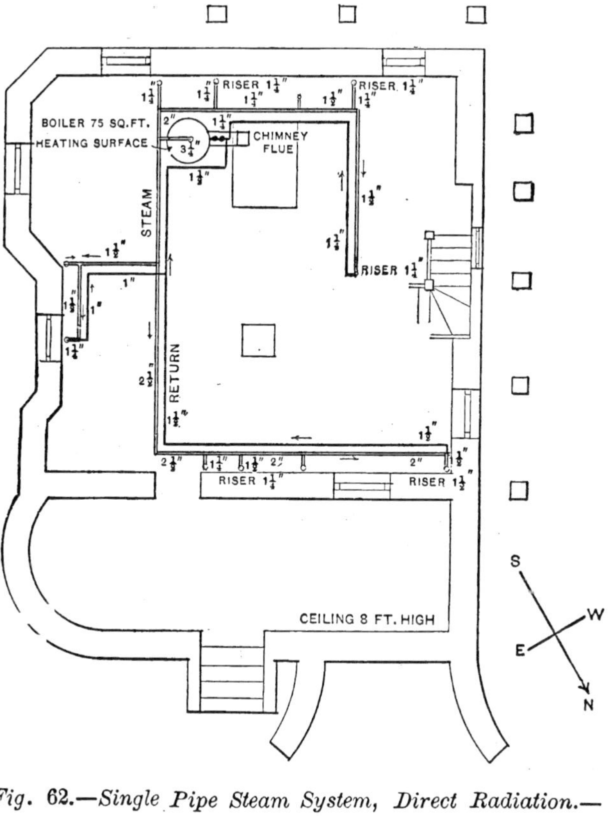

Fig. 62.—*Single Pipe Steam System, Direct Radiation.—Basement Plan.*

second floor plans of house heated by single pipe steam direct radiation)? How arranged? What are its advantages? Has the system been much used? If not, why?

Answer. A single or one-pipe system of steam heating, as known to the trade, is where one pipe performs the duties of both the steam supply and return for water of condensation. The main steam supply from the boiler is

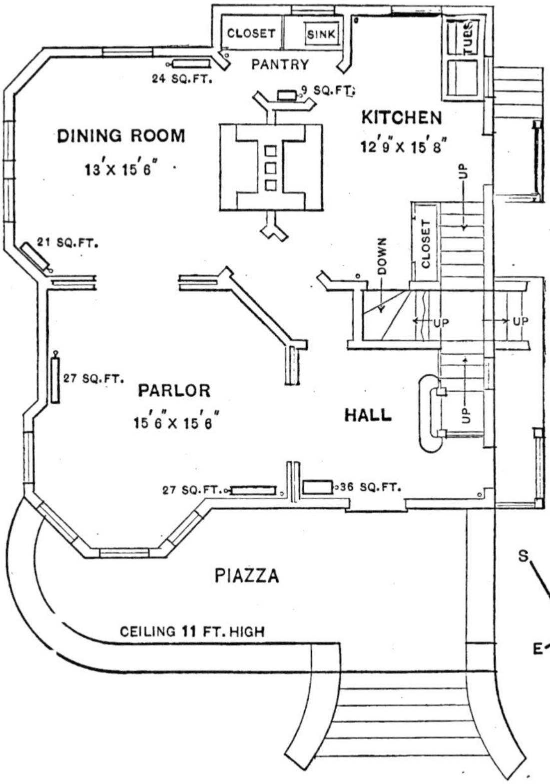

Fig. 63.—Single Pipe Steam System, Direct Radiation.—First-Floor Plan.

run in about the same manner as for the two-pipe system, of ample dimensions and inclination, with as few reductions in size as possible, connected up by eccentric tees and reducers when required.

The outlets for risers and first-floor radiator connections

are left as usual, the main being run to the furthest point to be supplied; there a relief, or return water pipe, is connected and run back with ample inclination to and connected up with the boiler with globe and check

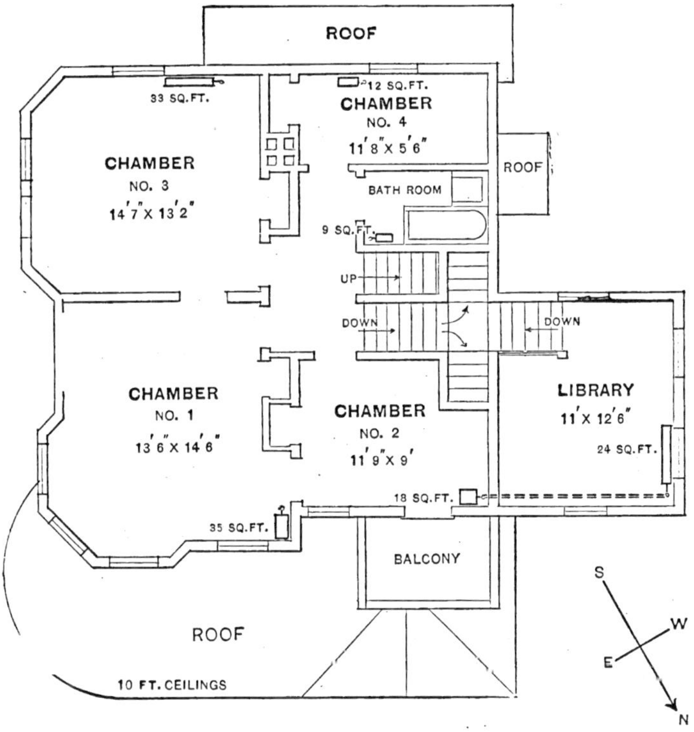

Fig. 64.—Single Pipe Steam System, Direct Radiation.—Second-Floor Plan.

valves, pass bys to sewers, etc. If the main pipe branches run in opposite or two or more different directions the same mode of connections, etc., is carried out. The branch pipes to the risers are taken out of the top of the main and connected up as usual. The branch connections from the risers to the radiators are made with only one side of the radiator. The valves used are either gate, corner angle or opposite angle, and so placed as to secure a free passage for both steam and water. Air valves are generally used in this mode of connection.

The steam passes from the boiler through the main pipe, the various branches and connections, to the first floor, and through the risers and branches to the upper floors, and the radiators, coils, etc. (Figs. 65 and 66). The inclination of these branches is downward from the radiators to the risers and from the risers to the steam main, and the main is inclined downward to the furthest point. The water of condensation therefore flows from point to point and is finally taken out by the relief or return water pipe at the end of main and carried back into the boiler.

This particular arrangement is often used for small buildings of two or three stories high, such as cottages, small flats, etc. A modification of this system is often used for larger buildings. In this case a regular steam and return main are run parallel to each other in the cellar; the return water from the bottom of the risers is taken off by a separate pipe into the return main. The reliefs to the branches from the first floor radiator supply pipes are treated in the same way. Of course the inclination of all these branches is from the main supply pipe down toward these relief points, thus promptly clearing the pipes of all water formed in them or received from above.

The advantage of this system is its extreme simplicity in construction and management (there being but one valve on each radiator). This greatly reduces the first cost, and when properly put up perfect circulation and noiseless working are assured. It can be adapted to all the various systems of steam heating—direct, indirect, or direct-indirect.

In the early days of steam heating this system was much used for low pressures, but from imperfect design and proportions and poor workmanship it was gradually abandoned. When high pressures came into use the two-pipe system was almost entirely used, but for low pressure work, in its improved form, the single pipe is well suited.

Q. 2. How many different forms of the single pipe system are commonly used, and what is each style best adapted to?

A. The systems which have been used may be distinguished as one which receives the steam supply from below,

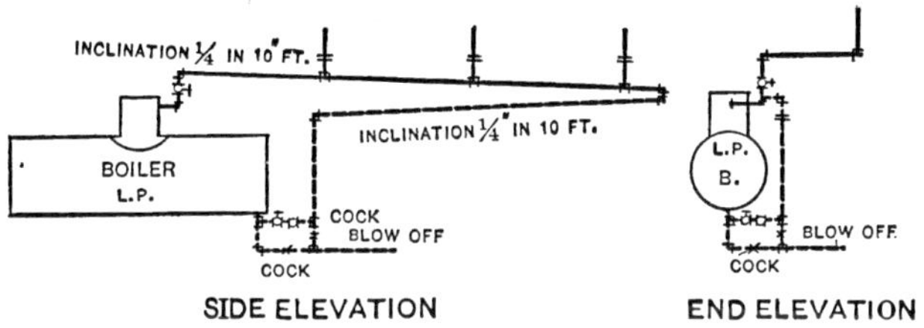

Fig. 65.—*Steam, One-Pipe System.*

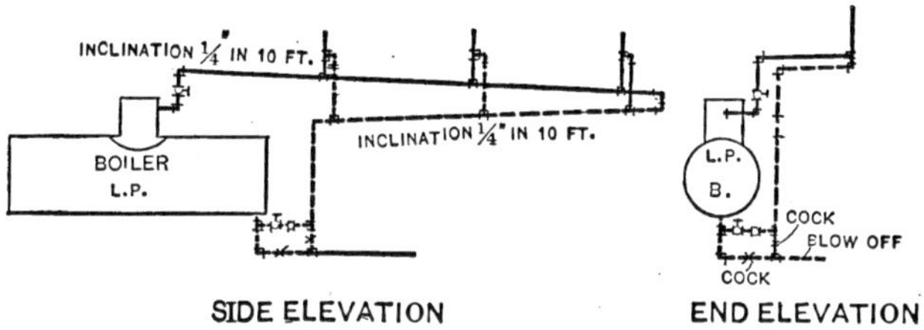

Fig. 66 —*Steam, One-Pipe System, with Risers Dripped Into Return.*

the other from above the first floor, The former has just been fully described; the latter may be briefly stated as the reverse of the other, except that the return water mains only are run in the cellar, as in other systems. The steam is taken from the boiler in a large steam main or riser, and is carried direct to the upper floor or garret; there it is subdivided and carried around to the points where the risers come up from below; they are then connected up with valves, etc. (Fig. 67). These risers pass down through the various floors, with outlets for branch pipes on each floor. At these points they are connected up with the radiators with one valve, as before described. The supply of steam passes into each radiator as it descends in the riser and the water of condensation flows down the riser with the steam until it reaches the cellar, where it passes through the connecting pipes into the main return, and from there into the boiler. Where there is ample room on the top floor this

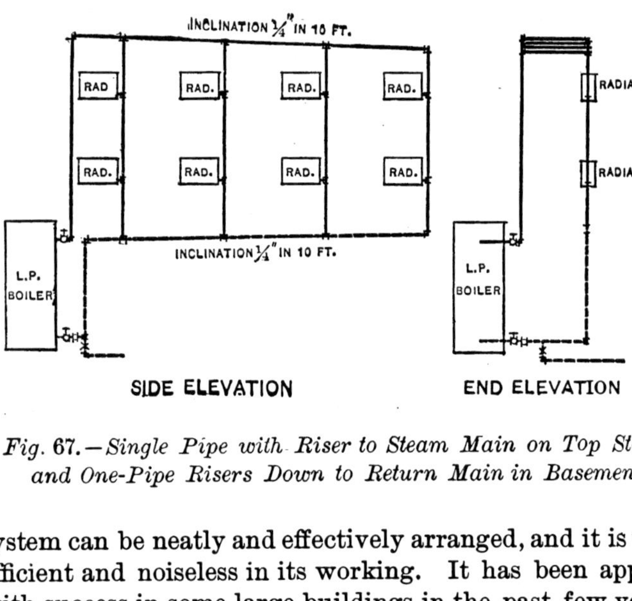

Fig. 67.—Single Pipe with Riser to Steam Main on Top Story, and One-Pipe Risers Down to Return Main in Basement

system can be neatly and effectively arranged, and it is very efficient and noiseless in its working. It has been applied with success in some large buildings in the past few years. It has the advantage of requiring only one valve on the radiator, but some object to the use of the upper story and the running of the large riser as unnecessary and also to locating the means of controlling the risers too far from the boiler room, in the attic.

Q. 3. Is there any difference in the construction of the boiler, the number or arrangement of parts or attachments used in connection with the boiler, etc.?

A. No differences of construction in the boiler, attachments or connections are required. They are the same as for any other low pressure steam system.

Q. 4. Is there any difference in the sizes and arrangements of the steam supply and return mains? If so, in what respect?

A. The steam mains, having two duties to perform, being conduits for both steam and return water, are larger, and having a greater amount of water in them than in the other system are made of generous proportions and run large size to their ends. The return mains are of moderate size, allowing for the friction of the water in a long run, and are of about the same size from end to end.

Q. 5. Is there any difference in the number, size, mode of running and connecting up of the risers? If so, in what respect do they differ from other systems?

A. There are no differences in the mode of running the risers and making connections other than those already described; there being but one riser and connecting branch to radiator, only one valve is required; the same is the case all through.

Q. 6. Are the radiators or coils used of the same make and arrangement? If not, in what respect do they differ?

A. The radiators and coils generally used in these systems are the same as in other systems, there being but one pipe connection with the radiator; this, however, is of larger size than in the other systems and the opening on the other end of the radiator is plugged up.

Q. 7. Are simple hand or automatic air valves and pipes required for this system? How are they usually arranged and operated? Have any other modes of arrangement or operating, without air valves, been tried? What are they?

A. The hand or automatic air valves are of the same make commonly used, but they are placed in the last pipe on the end opposite to where the steam is taken in. In some cases effective work has been reported without air valves, as where the steam supply comes from above. But it is safer and better practice to have them on the radiators.

Q. 8. What kind of valves are used in connecting up the radiators? Which are considered the best?

A. The valves used in the connections to the radiators are the same as in other systems—gate, corner, angle and opposite angle are generally used, and have proved very satisfactory.

CHAPTER VI.

INDIRECT STEAM HEATING.

Question 1. Is there any difference made in the boiler and attachments or the running of the main steam and return pipes when used for indirect heating instead of direct? If so, what are they, and why?

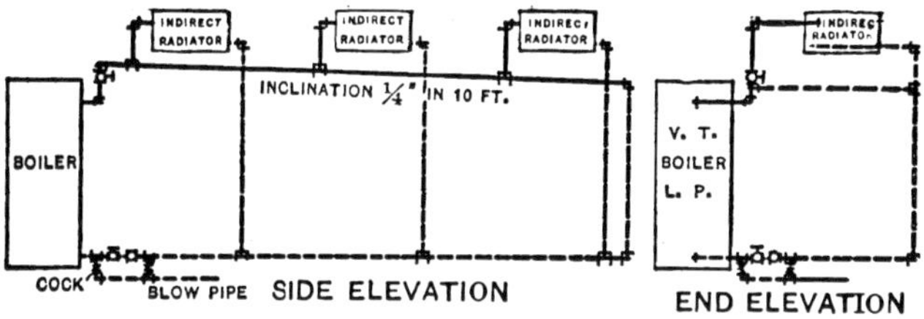

Fig. 68.—*Indirect Steam, Return on Floor.*

Answer. The boilers and attachments for indirect steam heating are in every way the same as for direct heating. The main steam pipes are the same, but the return mains are sometimes run along the wall near the floor, or in covered trenches beneath the floor. (See Fig. 68.) This is sometimes necessary, as, the indirect radiators being suspended from the cellar beams, the return pipes come down too low to be run with the steam mains, as they would obstruct the head room in the cellar.

Q. 2. What kind of radiators are usually used for indirect heating? How are they made and connected up? Where located and how supported and operated? Where are the air valves placed?

A. The radiators used in indirect steam heating are of many forms and patterns. Those which were first used were of the box coil style and generally made of 1-inch wrought iron pipe; subsequently flat cast iron sections with projections or pins on their sides were used. This was to increase the amount of heating surface and break up the air currents so as to bring all parts of the air in contact with the heated surface of the iron sections.

There are many other forms used, some with fins or extended surface, as it is called, but the extended surface can only be estimated as of 10 to 15 per cent. of the value of prime surface with which the steam is in direct and close contact. The steam connections are made on the top at one end with both the box coils and cast iron sections and the return connections on the opposite side on the lower end of coil or stack.

The radiating surface is best located as near the uptake or hot air flue as possible, and with few if any turns or corners, with easy curves in pipes, and as free from obstructions or deflections as practicable in all connecting pipes. Regard should be also had to the fresh air supply and to the direction from which the prevailing winds in winter blow. The coils or sections are supported by eye bolts or other hangers made fast to the beams above the radiators, leaving ample space between them and the top of the inclosing boxes. When the radiators are located in brick chambers they are supported by pieces of pipe or iron bars resting in the brick work. The cold air supply enters below the radiators, passes upward between and around them and collects in the space above them, then goes to the uptake or hot air flue to the rooms to be heated.

The air valves are placed near to and connected with the section or pipe nearest the return pipe.

Q. 3. Of what are inclosing boxes or chambers usually made, and how constructed?

A. Inclosing boxes or chambers were originally made of brick, built up around the coils and running from the floor of cellar to the ceiling. Heating coils are now usually inclosed in wooden boxes made of narrow matched boards lined with tin or galvanized sheet iron. Sometimes they are made of two thicknesses of board with heavy sized paper placed between and laid so as to break joints and prevent the leakage in of dust and impure air from the cellar. In other cases both ducts and boxes are made of galvanized iron with soldered joints, or they may be made with bolted joints with some elastic substance between. In all cases either air tight spaces or some non-conducting material should inclose the galvanized iron boxes so as to prevent the loss of heat in the cellar.

Q. 4. Of what materials are the cold air supply ducts usually made? From whence is the air taken and how is the supply controlled?

A. The cold air supply for radiators is usually taken from some window or opening in the wall made for the purpose on the north or west side of the building, as the prevailing winds are from that quarter in winter.

The quantity of air admitted is regulated by a pivoted damper, operated by hand, or automatically by a steam pressure damper regulator. As the steam pressure increases or decreases in the boiler the damper is opened more or less or closed entirely.

Other devices have been used for this and kindred purposes, to control the temperature of the air admitted to the rooms, etc., but all operate on the same general principles.

Q. 5. How is the air, after being heated, distributed to the various points where needed?

A. The air after being heated by the radiators passes up through tin or galvanized iron ducts placed in the wall when the house is being built, or run up outside of the inner side of wall and inclosed by fire proofing, or boxed in with wood with inclosed air space or mineral wool filling around it.

The best practice is to run a separate hot air duct to each room, with a register of ample size at the top. Ordinarily this is not practicable, so one flue of increased size with deflecting plate or damper in it, also a register on each floor, is made to answer the purpose.

Sometimes valves or dampers are arranged in the cellar where the cold air ducts are located and so connected that cold air may be introduced into the hot air flue and the air tempered to any degree desirable. This can be done by rods and levers operated from the rooms above, or automatically by special devices.

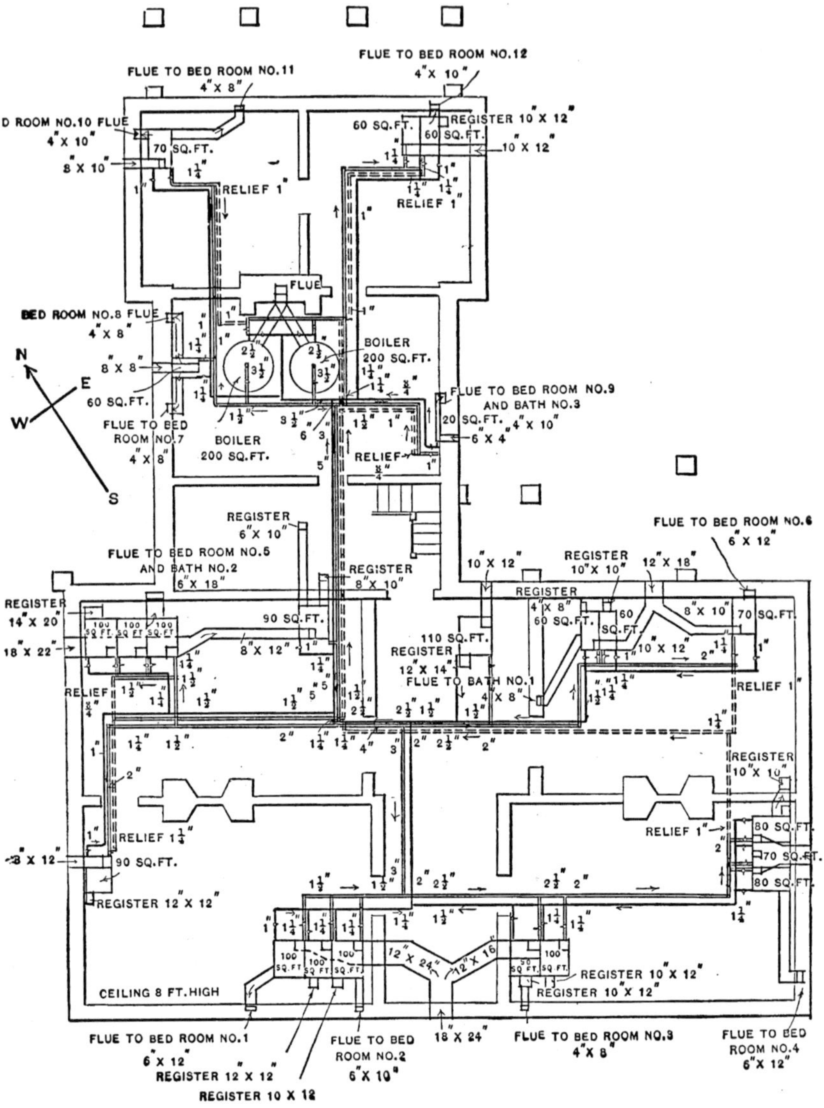

Fig. 69.—Steam Indirect Radiation, Two-Pipe System.—Basement Plan.

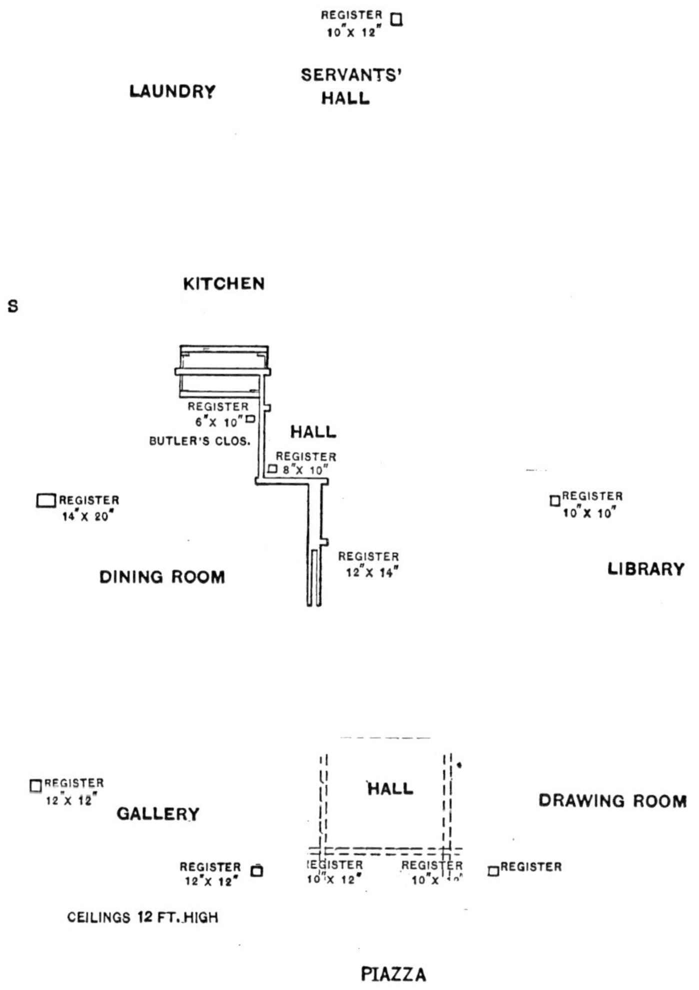

Fig. 70.—Steam Indirect Radiation, Two-Pipe System.—First-Floor Plan.

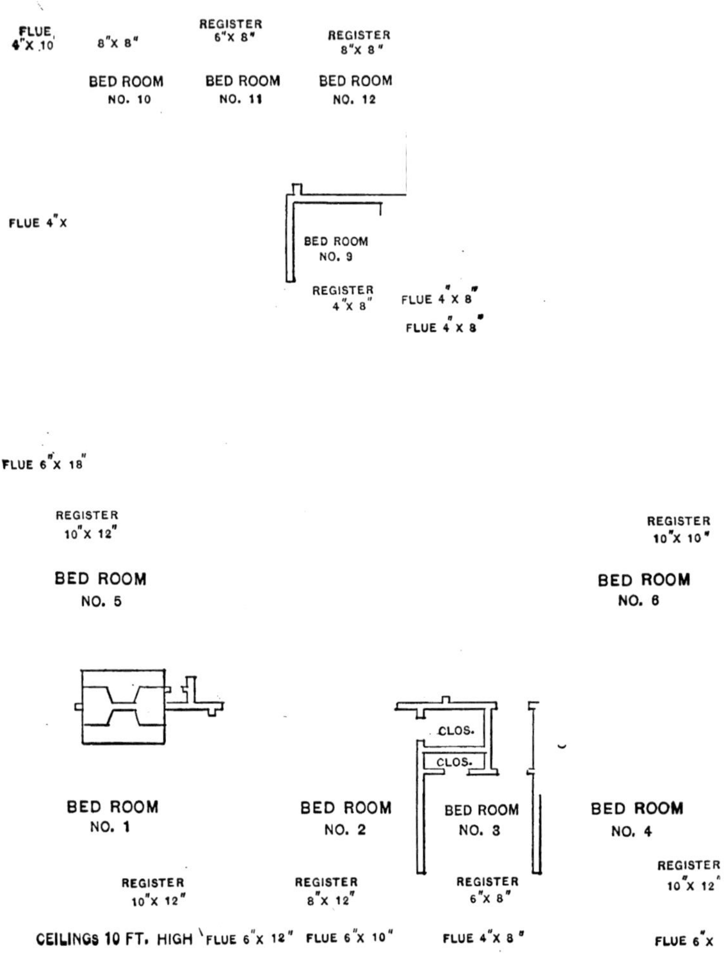

Fig. 71.—Steam Indirect Radiation, Two-Pipe System.—Second-Floor Plan.

Q. 6. Must proper means for exit of air as well as inlet be always provided when the indirect mode of heating is used? What is it usually called?

A. Proper means must be provided for the exit of the cooler and denser air in a room before the warmer and lighter air in the hot air flue can enter. Ventilation is accomplished in various ways—an open fire place, a fanlight over the door into the hall, even a partly opened window, is resorted to, but all are in a measure imperfect. The best practice is to construct flues for ventilation of ample size, lined with tin, galvanized sheet iron or glazed pipe, in the inner walls, well protected by non-conducting material. Opening into these are always placed, near the top and sometimes near the floor of a room, registers, with cords attached, so that either the top or bottom ones may be opened or closed. To quicken the action of these air currents sometimes aspirating coils made of steam pipes are placed in the flues at the bottom or near the top so as to assist the exit circulation. Steam coils are sometimes placed beneath large ventilators located in the roof over the staircases and halls. The above described system is one of the best and simplest modes of heating, at the same time supplying fresh air to the buildings, or heating and ventilating, as it is called. (A two-pipe indirect system is illustrated in Figs. 69, 70 and 71.)

Q. 7. May the two systems of direct and indirect heating be used in the same building in combination?

A. The two systems of direct and indirect heating, by steam, are often most effectively combined in the same building, the indirect being used for the first floor and the direct for the upper floors and the more inaccessible parts of the lower floor, thus securing an ample supply of warm fresh air to the house, which in moderate weather is sufficient to heat it by the warm air rising from the lower to the upper floors, the direct system being used only in very cold weather or when any room is closed and isolated from the halls and lower floors of the house. (A combination one-pipe steam system is illustrated in Figs. 72, 73 and 74.)

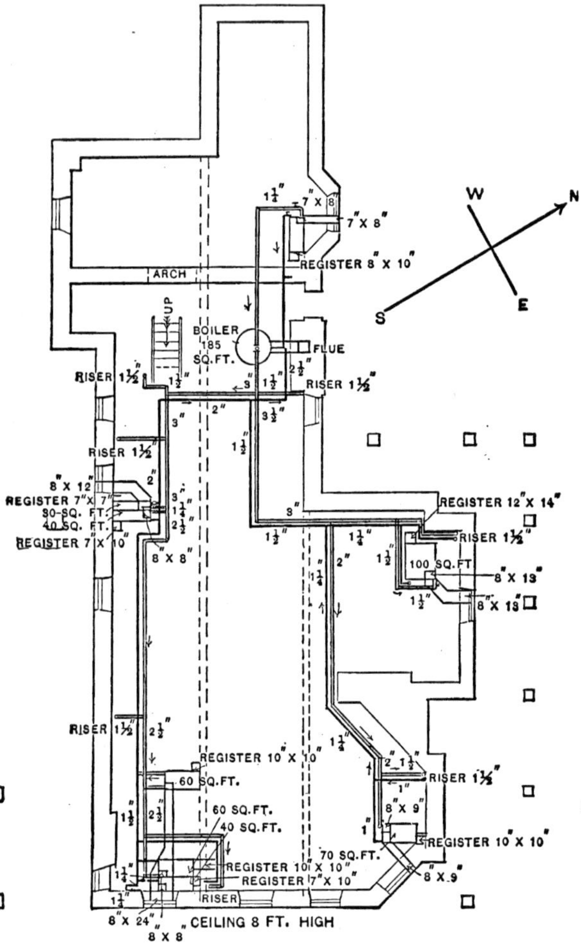

Fig. 72.—Steam Direct and Indirect One-Pipe System.—Basement Plan.

Q. 8. Are there any combined systems of heating houses by heated air differing from that mentioned above? If so, in what respect do they differ, and how do they compare in efficiency and economy?

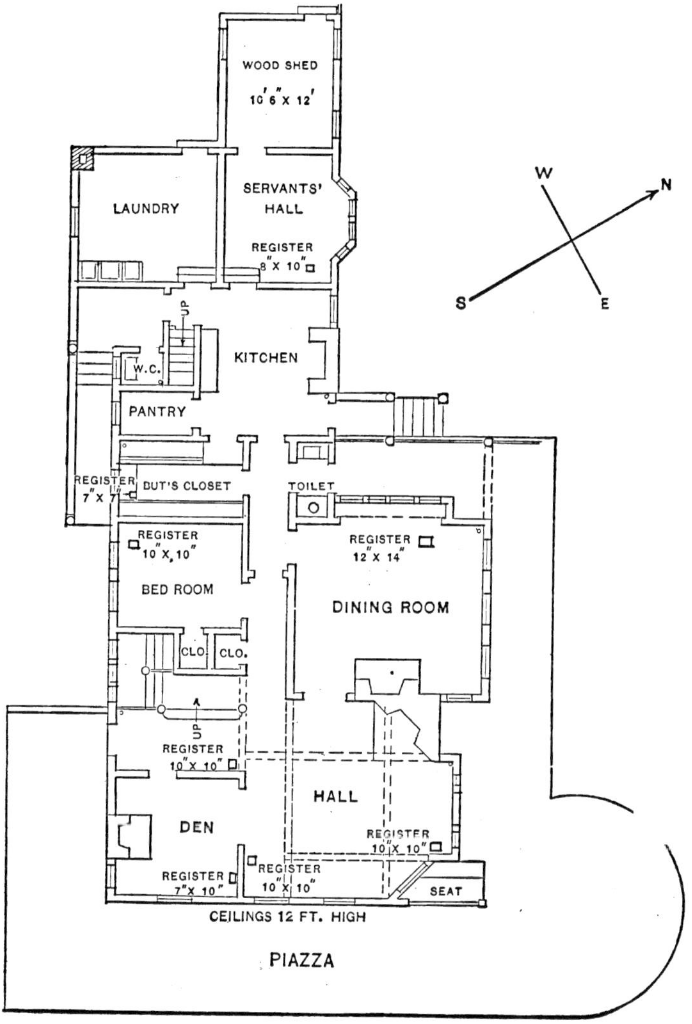

Fig. 73 —Steam Direct and Indirect One Pipe System.—First-Floor Plan.

A. There are other systems for heating houses, among them one called direct-indirect, a combination of some of the good points in both of the other systems, such as fresh

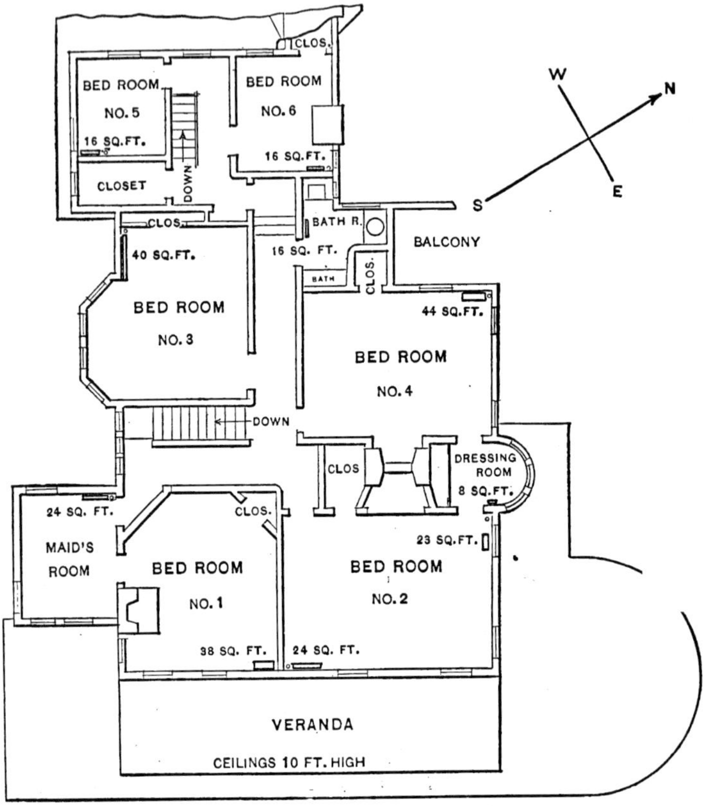

Fig. 74.—Steam Direct and Indirect One-Pipe System.—Second-Floor Plan.

warmed air combined with direct radiation in each room, all being under the immediate control of the person occupying the room.

The radiator is usually placed beneath a window or near it and the fresh air is brought in by a special duct from the outside of the house to a point near the radiator and has a controlling valve in it, or the air is admitted by a special construction of the window sash, or frame, or by other simple means. This system is usually less in first cost than the indirect and very efficient when properly designed, constructed and managed.

CHAPTER VII.

HOT WATER HEATING.

Question 1. In planning and proportioning a self regulating hot water heating apparatus for dwellings and build-

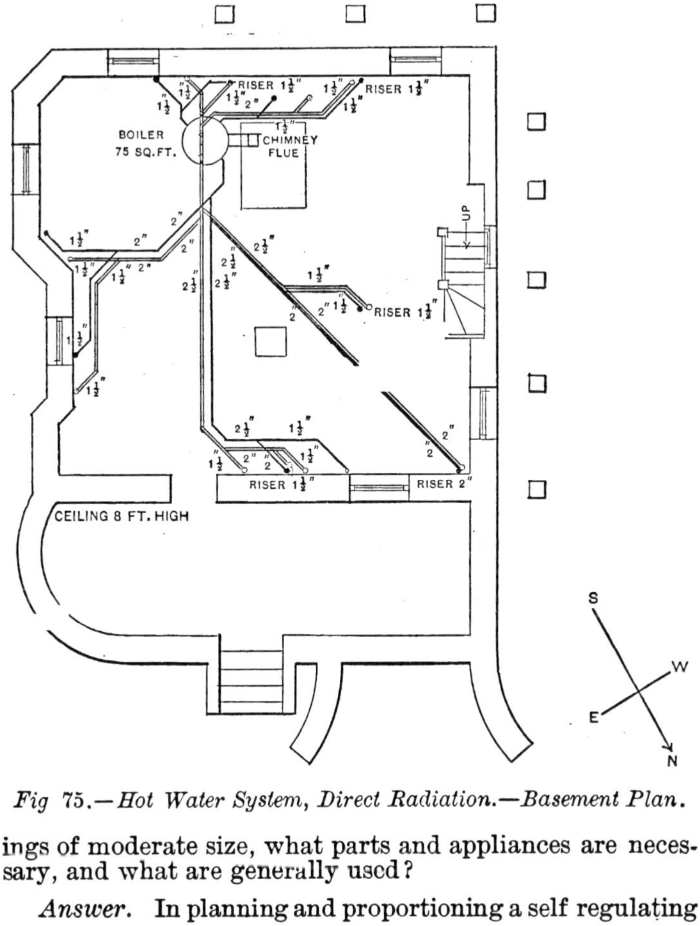

Fig 75.—*Hot Water System, Direct Radiation.—Basement Plan.*

ings of moderate size, what parts and appliances are necessary, and what are generally used?

Answer. In planning and proportioning a self regulating

hot water apparatus for dwellings and buildings of moderate size, the parts which are necessary and desirable are a boiler, main supply and return flow pipes, risers and return branches, supply pipes to the radiators and the returns with valves in one, an expansion tank and its trimmings, such as

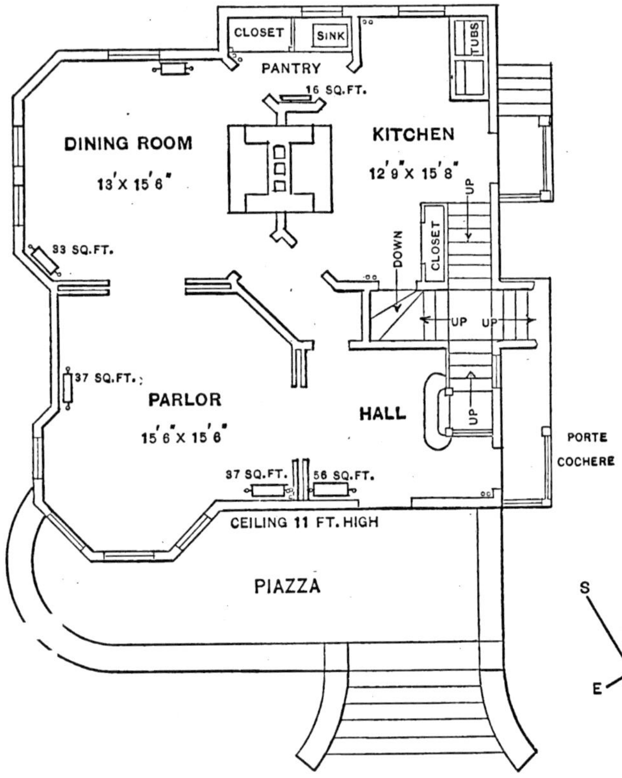

Fig. 76.—Hot Water System, Direct Radiation.—First Floor Plan.

a glass gauge, and sometimes an automatic water feeder to keep the water up to a proper level in case of leakage in pipes ; overflow pipe, etc.; also such appliances as are necessary to operate the boiler. A hot water direct radiation system is illustrated in Figs. 75, 76 and 77.

Q. 2. How many different systems of hot water heating apparatus are used? How do they differ?

A. There are two systems in use, the open and closed, or what may be called the low and the high pressure.

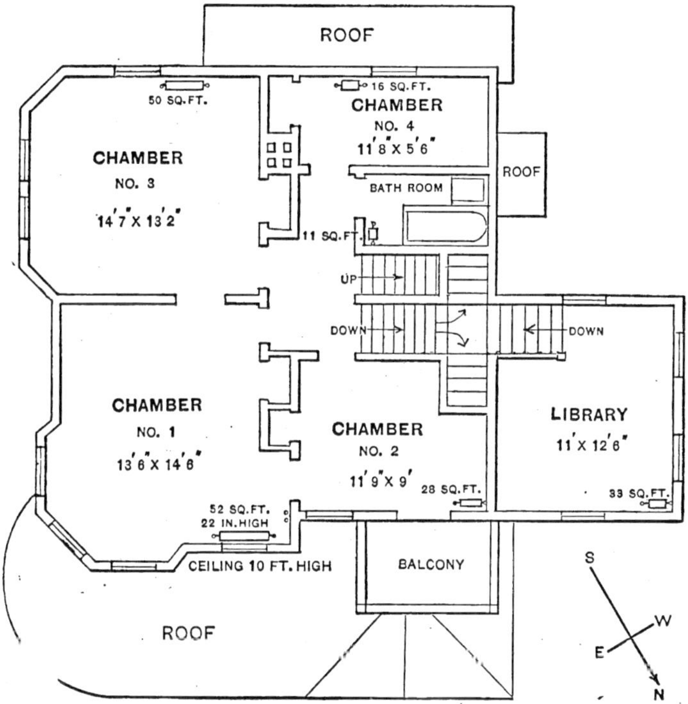

Fig. 77. Hot Water System, Direct Radiation.—Second-Floor Plan.

Q. 3. How does the boiler used in hot water heating differ, if at all, from the boiler used in steam heating?

A. All kinds of boilers are used for hot water heating. Some are of the shell style, some a combination of cast and wrought iron pipes, and others are entirely of cast iron. There is but little if any difference between the boilers used for steam heating and hot water heating, although many makers have a separate style for each purpose. In the steam boiler, when in use, the water is carried only to a certain hight, while in the hot water boiler the boiler and all the pipes above it are filled with water.

Q. 4. Are the attachments the same? If not, what are omitted or added, and why?

A. All of the attachments used on a boiler for steam heating are unnecessary when it is used for hot water heating. Sometimes valves are placed in the main supply and return pipes, but ordinarily they are not used. A blow off pipe to the sewer with a valve in it is necessary.

Q. 5. How are the main supply and return pipes run from and to the boiler? Are the proportions the same as for steam? If not, in what respect do they differ? How and where are the return pipes connected with the boiler?

A. The main supply hot water flow pipes are run out from the top of the boiler with a gentle rising inclination of about ¼ inch to 10 feet or more; the incline for the return pipe is the same as that of the supply and toward the boiler. (See Fig. 78).

The sizes of the pipes are made larger than those used for steam, and the capacity of the supply and return pipes is the same. All branches are run full, with as few reductions in size as possible. The supply and return are of the same diameter. The return pipes are always connected with the lower part of the boiler.

Q. 6. Is the inclination given to the mains all one way, or are they different?

A. The inclination of the mains is always down toward the boiler in a hot water apparatus; in the riser branches toward the mains, in the radiator branches toward the risers; in other words, all drainage must be toward the boiler.

Q. 7. Are separate mains ever run from the boiler to each rising line of pipes, or are the supply pipes to these risers taken from one or more main pipes, or in both ways?

A. Separate main supply and return pipes are often run from the boiler direct and back to it, particularly by makers of sectional cast iron boilers. This system virtually makes each line of pipe and its boiler section a separate apparatus, but it also makes an expensive arrangement. The main supply pipes are usually run out from the top of the boiler, as with steam, and the branches to risers and radiators are taken out, and their returns connected with the main return pipe to the boiler, thus making a reliable and efficient apparatus. In some cases part of the system

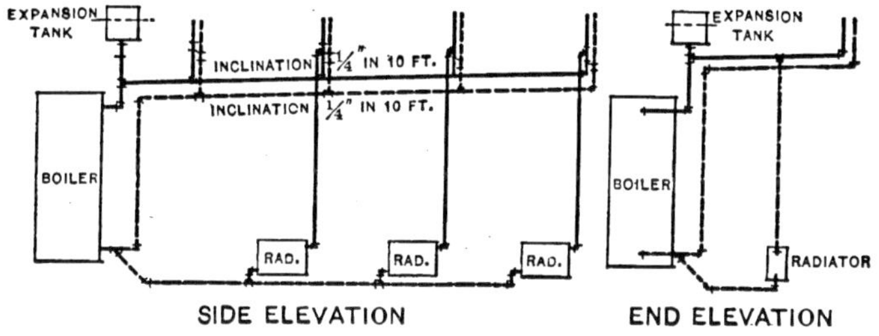

Fig. 78.—Hot Water System with Radiators on Boiler Level.

is supplied direct from the boiler, the rest from the mains, as the case may require.

Q. 8. How are the rising lines and their returns run?

A. The rising lines are run up from the cellar to the highest point where heating is required, the same as with steam. Sometimes the supply riser and return are run in channels in the walls or outside, together or separately, as may seem best. They are of the same size and the connections to and from the radiators and their controlling valves are alike.

Q. 9. Should a special stand pipe be run up to the expansion tank, or may a line of risers and returns be used for that purpose, and how are they arranged?

A. Special lines are sometimes run up to and connected

with the expansion tank, but a connection made with one of the risers and returns answers the purpose and secures constant circulation through the tank, thus preventing freezing when placed in an exposed position. Proper arrangements should be made for the expansion of risers and the pipes connecting with the tank.

Q. 10. What is an expansion tank? Where is it placed? What parts are attached to it, and for what purposes?

A. An expansion tank is a water tank or pot, placed in a position several feet above the highest radiator or coil to be supplied with hot water. It is made of galvanized sheet steel or iron or of cast iron, with an opening for the connections to rising pipe in or near the bottom; an opening for an overflow and vent pipe at the top; an opening in the side above the centre for the connections for the water feed supply; a glass water gauge connected up on the side of it to show the hight of water within it, and sometimes a ball and cock water feed regulator and attachments working automatically. Where there is no water works supply at hand a funnel is placed in its top so that it may be filled by hand when necessary, or it is pumped up from below. Other forms of tanks have been and are now used. Automatic gauge indicators are made, which, when placed and connected with the pipes of the system and located near the boiler, show the hight of water in the tank by the weight of the column of water in the pipes.

Q. 11. Why is an expansion tank or pot used? Are they always used?

A. Expansion tanks or pots are used with all styles of hot water apparatus, whether of the open or closed circuit construction, low or high pressure. The action of an expansion tank in a measure is like that of a safety valve on a steam boiler; it allows the apparatus to relieve itself and prevents overpressure. With the closed system the air in the upper part of the tank is compressed when the expansion of the water in the boiler, pipes, etc. is greatly increased by heat, and acts as a cushion or spring.

Q. 12. What kind of radiators are used for hot water heating? Do they differ from those used for steam heating? If so, in what respects?

A. The radiators used for hot water heating are of the same general form and make as for steam. The pipes and sections, however, are connected both at the top and bottom.

Q. 13. How are they connected with the risers? How many valves are used on each radiator? How are the branch pipes run, and where are they connected with the radiators?

A. They are connected in a similar manner to steam with the risers, but one valve is generally used, placed in the supply or return branch connection pipe, as may be most convenient.

The branch pipes from the riser to the radiator are usually connected with the bottom of the radiator, but under certain conditions, as when on the same level with the boiler, it has been found better to connect the supply pipe with the top of the radiator on one end of the radiator and the return from the bottom of the opposite end.

Q. 14. What kind of valves are used on hot water radiators? Do they differ from those used for steam? If so, in what respect?

A. The only valves used are on the radiators, with a blow off cock on the boiler. The valves used are gate, angle, corner, opposite angle, or any form which will give a free waterway. Sometimes a quick opening or plug valve is used.

Q 15. Are air valves used on hot water radiators? If so, why and under what conditions?

A. Air valves are generally used on hot water radiators where they are connected up at the bottom. The make of valves usually employed are what are called key air valves.

These are opened only when the apparatus is newly filled and started up. When the connections are made with the top and bottom of the opposite ends of the radiator no air valve is required.

Q. 16. Are there any disadvantages attendant upon the use of a hot water apparatus, direct or indirect? If so, why? How can they be avoided?

A. The principal objections to the use of a hot water apparatus are its high first cost, want of quick response when more heat is required, inability to be automatically controlled, its liability to freeze up and cause serious damage to the house and furnishings, and the length of time that is required for a radiator to cool after it is shut off. It, however, has its advantages. Careful handling and attention reduce the risks and increase the comfort it gives by a uniform and healthful temperature.

Q. 17. How should the pipes and iron work in the cellar be finished? And the radiators and exposed pipes above the cellar?

A. The pipe and iron work in the cellar should be finished with one or two coats of black varnish well laid on, except the bright work around the boiler and the indirect radiators. The radiators and exposed pipes above the cellar should be painted or bronzed either silver or gold, as may be desired.

Q. 18. Is the felting or covering of steam or hot water pipes advantageous? Should risers in the walls be protected, and how best done?

A. The felting or covering of either steam or hot water pipes as a preventive to the radiation of heat from pipes is very advantageous, particularly where the heat is not needed in the cellar, and also to protect them from freezing when run in exposed walls or places. Pure water of condensation from steam or boiled water will freeze much more readily than ordinary or impure water. It is often necessary to leave the pipes uncovered in the cellar to protect the flooring of the floor above, as a cold floor invariably makes an uncomfortable room, so it is no loss when the heat is so utilized. The pipes in the cellars of dwellings should be run large with this object in view.

Risers in the outer walls should always be protected by felting, mineral wool, wooden boards, or some other good non-conducting material.

Q. 19. Is there more than one system or manner of running pipes and operating hot water heating? If so, how is it arranged? Can the risk of the water freezing in the pipes be obviated or guarded against? If so, by what means?

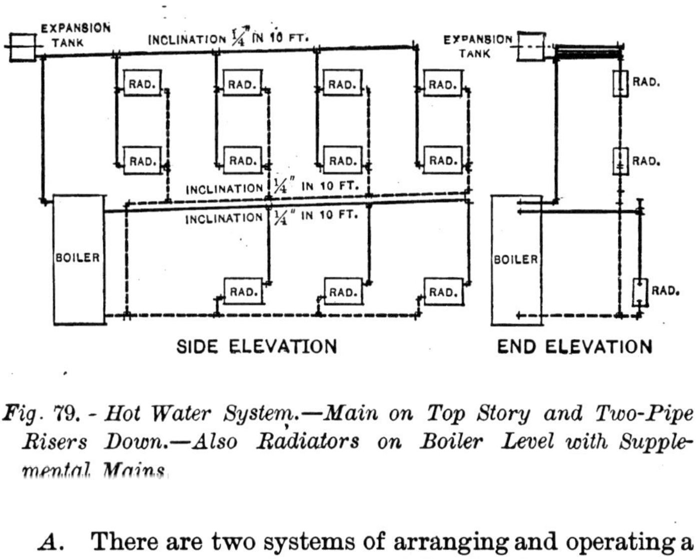

Fig. 79. - Hot Water System.—Main on Top Story and Two-Pipe Risers Down.—Also Radiators on Boiler Level with Supplemental Mains.

A. There are two systems of arranging and operating a hot water apparatus—the open and the closed circuit, as before stated. In the former an open expansion tank is used; in the latter the expansion tank is closed and the whole apparatus works under pressure, more or less, as the situation requires (Fig. 79). Provision should always be made for drawing off the water in the pipes during cold weather when the apparatus is not in use and also during the summer season.

Q. 20. How are railroad cars, offices, etc., on the same floor and level with the boiler heated by hot water? Are there any risks and disadvantages connected with the system? Is it safe and efficient?

A. Railroad cars, offices and apartments have been and are now frequently heated when the generator or boiler is on the same level as the radiators or heating coils, and in some cases when the radiating surface is below the boiler. In a generator or boiler made up of a coil or coils of pipe or other form and placed in a sheet iron shell or casing lined with fire brick, the main flow pipe is taken out of the top of the generator and often dropped down to the coil or radiator near the floor and the connection made; or it may be carried overhead to some distant point, or along the floor, and connected up where desirable, the return main running along or below the floor back to and connected up with the boiler. The expansion tank is located near to and above the generator, and in the case of cars is placed above the roof in a closed box. Circulation coils of iron pipe 1½ to 2 inches diameter are usually used for such purposes.

To prevent the water from freezing in certain cases, as in isolated offices, railroad cars, etc., when not in operation, a saturated solution of salt in water is used. This form of apparatus is run upon the closed circuit or high pressure basis. With proper care in construction and management the risk is small, as the apparatus is tested to a much higher point than it is ever likely to go to. Its efficiency has been fully proven by many years of varied service. The parts are made chiefly of extra heavy small size wrought iron pipe; the expansion tank of the best gun metal.

CHAPTER VIII.

SINGLE PIPE MAIN SYSTEM.

Q. Has a single pipe main system of hot water heating ever been used? How is it arranged and constructed? How is the proper circulation of the water induced and maintained? What are its advantages, and how does it differ from other systems in use? Is it reliable and effective?

A. The single pipe main system has been and is now used for hot water heating (Fig. 80). A single main pipe, which answers the requirements of both supply and return,

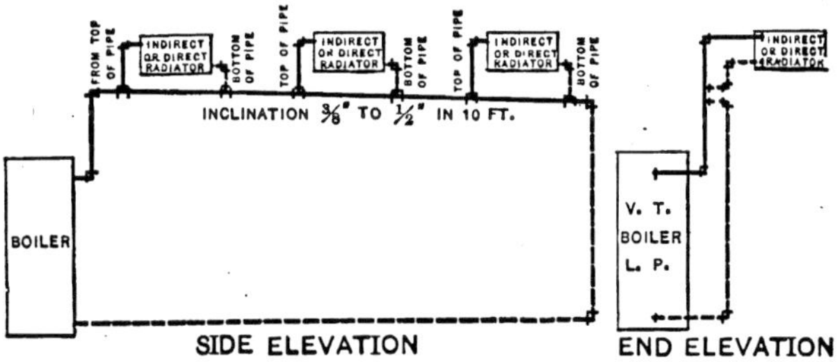

Fig. 80—*Hot Water One-Pipe Indirect System.—Supply Taken from the Top of Main, Return to the Bottom of Main.*

is run out from the top of the boiler—tubular, sectional, cast iron, or other form in common use. This pipe, which is of extra size, is run around the cellar near to the ceiling and is the starting point of rising lines and the branches to radiators on the floor above. After supplying all these points, as above stated, it is run directly back to and is connected with the bottom of the boiler. Its inclination from the top of the boiler back to the bottom of the boiler is sometimes as much as 1 inch in 10 or 12 feet, or as the length of run and arrangement of pipes will admit.

The branch supply pipes to the risers and radiators on the floor above are taken out of the top of the main pipe; the returns from the same source are carried back to and are connected up with the sides and, when the proper eccentric side outlet fittings can be had, as near to the bottom of the main as possible. All these branch pipes incline downward toward the main.

There are two risers run up from the cellar for each line of radiators—supply and return. These risers are connected at their lower ends to the top of the main for the supply, and to the side or near the bottom for the return, as described above, all of the same size. The radiators are connected to the risers by branch pipes of like diameter, and they incline from the radiator down to the risers. The radiators used are the same as for other hot water systems and only one valve is used, generally placed in the supply pipe. Air valves are employed as usual. No valves are used on the mains or around the boiler. A blow off pipe run from the boiler to the sewer, with a brass cock in it, is all that is necessary.

An open circuit expansion tank is placed at some point above the highest radiator, with all connections and trimmings required as for other systems of hot water heating.

The circulation is induced and maintained on the same principles as other systems, by the expansion and lightening of the water by heat. As the water in the boiler is heated it rises to the top and passes out into the main pipe, the hotter water being always on top. This passes out to the supply branches, the risers and radiators. As the heat passes out of the water through the radiators, etc., it becomes heavier, descends by the return pipes to the lower part of the main where the water is cooler, and flows back with it to the bottom of the boiler, there to be reheated and forced out again into general circulation.

The advantages claimed for this system are simplicity, less first cost, efficient working, etc. While it may appear that little can be saved by the substitution of one extra large pipe for two of smaller size, yet those who have had long experience say there is, and disinterested and competent judges have said that where the parts are properly proportioned and put up the circulation is perfect and the efficiency equal to any other system used for buildings of moderate size.

CHAPTER IX.

HIGH PRESSURE STEAM HEATING.

Question 1. Has high pressure steam with a direct return of the water of condensation from the radiating system back to the boiler been used? If so, in what manner and in what class of buildings was it in the past or is it now most used? What is the difference between the high and low pressure system?

Answer. High pressure steam has been used for many years—in fact, from the first introduction of heating by steam in this country—for special purposes, particularly where high pressure steam had to be carried in the boilers for driving steam engines in factories, and where steam and hydraulic elevators were used in stores, office buildings, and factories of all kinds, or where it had to be carried a long distance before it could be applied to the purposes desired. All such places necessitated the services of a competent engineer. Low pressure steam was and is used in dwellings and small buildings where no power for machinery is required, and where no specially skilled man is necessary. There is but little if any difference between the actual working of the high and low pressure direct return steam heating systems. Greater care is generally taken in the construction and erection of the former and increased strength is given in the various parts. The temperature of the steam used is the chief difference. The higher pressure having the higher temperature gives out more heat for a given amount of radiating surface.

Q. 2. How is the apparatus arranged and the parts necessary constructed?

A. The apparatus and parts required are about the same as for low pressure—a boiler of any form which will safely carry the pressure required, generally of the horizontal tubular, vertical tubular, of wrought iron or steel plates, also sectionals of various forms, chiefly made of steel or wrought iron; small pipes with drums, headers, etc.; regulating apparatus for draft, feeding, etc.; a two-pipe system of mains and returns; the same for risers and branches throughout. The radiators and coils are arranged in the same manner, and the connecting branches in all parts drip the same as for any two-pipe steam heating system. The connections to the boiler are the same.

Smaller sizes of pipe throughout can be used in the high pressure system than in the low pressure, and advantage is taken of this by many, and fairly good work may result, but in modern practice, for the best work, there is but little difference. A small saving in pipe, fittings, and valves may sometimes be effected, but often with a loss of efficiency.

Q. 3. Is there any difference in the construction of the boiler? If so, in what respect and why?

A. The boilers used are of the types above mentioned. Wherever sheets are used of either iron or steel, they are made thicker; the riveting in the horizontal seams are made double, triple, or more, depending on the pressure; the head plates are thicker, the braces are heavier, and all parts are more carefully put together.

Q. 4. What changes if any are there required in the parts around the boiler, in the attachments, regulating apparatus, etc.?

A. The parts connected with the boiler are used for like purposes as those in low pressure work, but of heavier make with some difference in form. The main steam heating supply, power, and return water valves, checks and other valves are of the best make. The water column, glass water gauge and connecting parts; also three gauge cocks of larger size and similar quality. A steam gauge, with registry of from 120 up to 200 pounds per square inch, or more, if necessary, is used. The steam damper regulator is much larger and more powerful, carrying a heavier weight on a longer lever; also a larger rubber diaphragm. The safety valve in general use is of much larger size, with a longer lever and heavier ball; sometimes safety valves have strong springs in place of a lever and ball to keep them down in place.

Q. 5. What difference is there in the manner of supplying water to the boiler?

A. The pressure of steam carried being too high generally for the feed water to the boiler to be run in from a tank, the supply from the city water works, or other source, as used for low pressure boilers, some other means must be used. A feed water injector, a steam pump, a return steam trap, or other device is generally used, and may be controlled directly by hand or automatically operated. The injector may be set so as to feed a limited amount of water constantly to the boiler. The steam pump may be similarly arranged or intermittently used. The return steam trap can be arranged so as not only to return the water of condensation from the heating apparatus of a building into the boiler, but to feed in any additional water necessary at the same time.

Q. 6. What is a feed water injector, the principle on which it acts, how constructed, arranged, properly placed, and connected up?

A. A feed water injector is an instrument used for forcing feed water into a boiler, like a steam or other driven pump, but its mode of operation is different. The escape of steam under 100 pounds pressure into a vacuum is at the rate of about 2000 feet per second, and when this is concentrated to a small point and it strikes cool water, it is condensed and imparts a certain portion of its velocity and all of its energy to the water. When this is properly concentrated and directed the water is forced along to whatever point required, as into a boiler. Cold water, or even at 140° can be used, but the warmer the water the less effective is the action of the steam, and above 140° the results are uncertain or even prevented altogether.

There are many patents on and makers of injectors. There are two kinds, single and double tube. The essential working parts are a steam inlet tube and nozzle and an outlet or water conveying tube or combining tube. The steam tube is generally conical outside, tapering toward the water tube; the inside is flaring with the large flare opening toward the water tube, or the reverse of the outside. The water tube is flaring toward the steam tube and a little larger than the outside of it. They are brought near to each other, leaving an annular space between them through which the water enters. A lever connected with the steam valve stem increases or diminishes the amount of steam admitted and correspondingly the amount of water forced into the boiler. There is a steam supply pipe, also a water supply pipe and a water discharge pipe, all properly connected. The working parts are inclosed in a brass case, and a water overflow pipe is connected at the bottom.

The double tube is a repetition of the parts of the single tube arrangement with some modifications. The lower is the sucking or water raising tube, and from it the water passes on to the second tube and by it is forced into the boiler. For a complete understanding of the injector a diagram showing sectional view must be examined.

The single tube was the first used and in that form was patented by a Mr. Giffard of Paris, France, about 1850. It was at first considered paradoxical and a toy. But further examination proved its value. The double tube is the most perfect for lifting purposes and was invented by Mr. Korting of Germany. It will lift water at 150° F., is not affected by variations in the steam pressure, and may be set to feed regular quantities for any length of time. Injectors will work with steam pressures varying from several hundred pounds per square inch down to the ordinary exhaust from a steam engine, and feed water to boilers carrying as wide a range of pressure as above named.

The location should be as near to the boiler as possible, and all the connections as straight as possible.

Q. 7. What is a steam pump? How constructed, located, set up and connected for feeding a boiler and general pumping purposes? What is the difference between a single and duplex or double cylinder steam pump? The advantages of each form?

A. A steam pump is a machine used for raising and forcing water from one level to another into open receiving tanks or closed tanks, with or without pressure in them, as with a boiler, etc. Independent steam pumps were originally made with one steam and one water cylinder placed on the opposite ends of a bed or frame work, the action being controlled by steam valves which were operated by an auxiliary piston, the latter being actuated by steam controlled by a small valve worked by tappets and adjustable arms attached to the rod connecting the steam and water pistons or plungers. Others were operated by the steam piston striking tappets in the steam cylinder at or near the end of each stroke, thus shifting the steam supply valves and reversing the motion. Others used a fly or balance wheel to carry the movement beyond the centers.

The duplex steam pump has, however, largely supplanted all the others. This is no more than two single pumps placed side by side on a bed, or the two steam cylinders and the two water cylinders are cast together. By a simple arrangement of arms and levers the motion of one pump, when near the end of the stroke, actuates the steam valves of the other pump, and the reverse, so that a quiet, easy and positive motion is obtained without dead centers.

The location of a pump in relation to the rest of the apparatus is often a matter of convenience. If used for boiler feeding it should be as near to the boiler as it can well be located so as to be within easy reach of the engineer, but should be protected from the dust from the boiler room. It should be so connected with the boiler that the steam supply may come direct; the water supply should be ample, and the discharge from the pump carried directly into the boiler or through a feed water heater to the boiler, or, if desired, so arranged that it may be used for the house water supply in emergencies. But there should be a separate pump for the house supply water, for when exhaust steam is used for heating, the oil from the engines and pumps passes through the pump with the return water of condensation from the pipes in the building before going into the boiler, and this affects unpleasantly any water for house use which might be passed through the same pump.

Q. 8. What is a hot water feed tank or feed water heater? How made, set up and connected up?

A. A hot water feed tank or feed water heater, as commonly used for heating feed water for the boiler, is made of sheets of iron or steel like a boiler, with heads of the same metal, well braced. In one of these heads there is a large man hole for cleaning out, etc. They vary in size from 30 inches diameter by 5 feet long, to 36 to 60 inches diameter by 6 to 12 feet long or more. Inside of this cylinder is placed a brass coil of 1¼-inch to 3-inch pipe or more. Where live steam is used for heating the water in the tank 1¼-inch pipe is usually used, but when the exhaust steam from the pumps and engines is used then the larger sizes of pipe are necessary.

The form of coil varies. With the larger pipes the two ends of the coil are passed through one head of the tank and the joint made up with lock nuts and packing. Sometimes only two pipes, the length of the tank, with a return bend or elbows and nipple connecting them on the far end, are used. Often four lengths or more are required. With the small pipe a top and bottom branch tee or header with center outlets are placed inside of the tank, and as many lengths of pipe, connected by return bends, as may be required, are used. The steam supply and return connections are made through the head of the tank, or through the top and bottom of it. This is a very simple and efficient form of heating tank.

Q. 9. What is a Berryman feed water heater, how made, set up and connected complete for work?

A. Later on the Berryman feed water heater, now so well known, came into use. This is really only the perfecting in mechanical details of the tank or heater already described. The cylindrical part is the same, and the brass tubes within are bent with an easy sweep to a U-shape, thus providing for expansion and obviating joints. In this form very light pipes can be used. These are expanded into a cast iron head which is bolted on to the wrought iron shell. There is another head below this, thus forming a divided chamber or compartment and leaving a space between the two heads. Across this is the division plate. One end of the tubes opens on one side of this plate, the other end on the other side, so when either the exhaust or live steam from the boiler comes into this space it passes into one end of the tubes, up through them, and down out through the other end. Thus the water which surrounds the tubes in the shell is very promptly heated and any sediment or impurities in the water fall to the bottom, where they are blown out. The connections are made with the exhaust pipe from pumps or engines at the bottom on one side, the discharge being on the opposite side of the heater. The water supply to the boiler is taken out near the top, the cold water supply is introduced near the bottom, the blow off from the top and bottom of the water shell; drips are taken from the bottom of the exhaust compartment.

Q. 10. What other classes of feed water heaters were first used and are now in general use? Their several advantages and defects?

A. Among the very first forms of heaters used (and some use it yet) was a coil of 1-inch or 1¼-inch pipe or larger, set in a cast iron casing, with holes in the top for the admission and release of the exhaust steam. A relief pipe for carrying off the water of condensation was taken out at the bottom. The cold water supply to the coil entered at the bottom and the discharge of hot water was at the top. The trouble with such heaters is that the amount of absorbing surface is so small that the moment the pump begins to work, the water then in the pipe soon passes out and the temperature of the feed water is rapidly reduced; besides all impurities held in it must go into the boiler, for if they remain in the tube they will soon close it up. Many other forms of heaters are used, which are modifications of the three types described above or have straight tubes, with expansion joints, and are very like an ordinary shell boiler.

Q. 11. What is a steam trap? How made, set up, and connected complete for work? How many different forms were or are now made, their special application, advantages and principles of action?

A. A common steam trap, as known in steam heating work, is an appliance used to relieve pipes and vessels containing steam from the water of condensation as formed, and at the same time to prevent the steam escaping.

There are many kinds; the action of some depend upon the differences in the expansion of metals; others the expansion of chemicals or the lighter hydro-carbons; others the force of gravity, as in the difference between the weight of air or steam and water.

Among the first used were those dependent upon the expansion and contraction of metals. They were called expansion steam traps, and were made with an outer case, generally of cast iron. Within this was a strip of brass curved to a certain shape, and made fast at both ends. In the center of this was placed a cap or valve, and below it a piece of pipe or an adjustable plug screwed into the casting. Sometimes this projection formed a part of the casting with a hole through it. This was the water outlet. The cap or valve rested upon this hole and closed it when steam came in the trap. When the water returned and cooled it the metal strip contracted and the valve was lifted off its seat and the water allowed to flow out. It acted very much like some of the air valves of the present time, such as the Davis and others of like make.

In others, like the Haws steam trap, the action is caused by the expansion and contraction of chemicals, in others of naphtha, or a combination with it of some other sensitive liquid, is used. Action is caused by expansion from extra heat. They run full open until the steam comes, then close quickly. These were good in their day, and a few are used now, but the metals become corroded and lose their sensitiveness in time and the chemicals or hydro carbons leak out, or in some way lose their sensitive qualities and reliability; the diaphragms also often break with their frequent bending and buckling.

Joseph Nason finally invented the steam trap known so long and favorably as the Nason steam trap. This is made with an outer casing or pot with a cover bolted on the top, having inflow and outflow channels near the rim or flange and on opposite sides. The channels which connect these with the interior are cast in the cover or upper part of pot. Inside of the pot there is a thin and very light cast iron open float with a valve on top attached to a stem in a tube. This float worked up and down as the pot was alternately filled with water of condensation or was discharged. There was a diaphragm inside of the casing which was attached to the cover of the trap. This served to direct the course of the water when it entered the casing. The float carrying the valve, when the case was entirely empty, was down. When the inlet valve was opened the water from the steam coils, &c., passed into the casing through the channel cast in the trap cover and fell on the deflecting plate; from there it passed into the casing around the float, lifting it up and closing the outflow valve. When the float itself became filled with water its buoyancy was lost and its weight of material caused it to descend, thus opening the discharge valve. When the float was emptied it lifted again and the action was repeated.

The stop valve, which is placed in the top of the cover, when open allows the water of condensation to be blown through directly to the receiving tank or the sewer without passing into the float or surrounding casing.

There are many different forms of bucket or float steam traps made, some with one, others with two valves, very similar to the low pressure feed water regulators, only they are applied to very different purposes, the one being used to control the feed water to a boiler, the other to control the outflow of water from the steam pipes, at the same time preventing the escape of steam. Robert Berryman, the inventor of a feed water heater, was among the first to utilize this form of steam trap, and this, with his other or gravity form of feed water regulator, led to the production of what is known as the "automatic steam return trap."

The steam trap is usually set up near the tank or pipe into which it is to discharge its contents, and is connected by an inflow and outflow pipe with a valve in each.

Q. 12. What is an automatic return steam trap? How made, set up, and connected complete for work? What are the principles and causes of its action, and to what system of steam heating can it be best applied? How many different general forms of steam return traps have been made?

A. The "automatic return steam trap" is a device by the use of which the water of condensation formed in steam heating pipes, or other sources of supply, is returned automatically, and with but little if any loss of heat, directly back to and into the boiler.

The two forms of this trap which first met with most public favor were the Blessing, or gravity trap, and the McNeill or bucket trap. The first relied solely upon the radiation of heat from the thin cast iron shell of trap for the densation of the equalizing steam; the latter used the exhaust system and discharged its contents immediately into the atmosphere. Both worked well, but the latter much quicker. As the heat contained in the equalizing steam had to be lost in either way, the quicker it was done the better. All of the return steam traps introduced since then have been modifications of these two systems, chiefly in the valve gear, but the stationary or bucket form in some shape is now almost universally used.

The "automatic return steam trap" was first produced in this country and introduced by James H. Blessing of Albany, N. Y., before 1873. A modified form of a like machine for the same purpose had been used in France some years before. The form of trap first used by Mr. Blessing was what he called his gravity trap. A cast iron frame work with two upright arms supported another frame and lever with knife edges at the side. This rested on the lower frame. On one end of this frame or lever hung a hollow cast iron sphere, and on the other end an adjustable counterbalance weight. To this sphere were attached two wrought iron pipes of different sizes, the lower or discharge pipe being one size larger than the upper or receiving pipe. These were run out from 8 feet to 12 feet to obtain easy spring of the pipe and motion for the sphere. The receiving or top pipe was connected with a receiving pot, into which flowed the water of condensation from the system of coils and radiators, and the discharge or bottom pipe was connected directly with the boiler, with check valve in each near the trap. A third or steam supply pipe of smaller size was taken directly from the dome of the boiler and connected with the valve case containing the steam equalizing trip valve. This case was connected with the top of the trap case or iron sphere. The whole apparatus was placed on the top of the boiler wall or on some shelf made for it, near to and above the boiler, so as to place the bottom of the trap case, when filled and down, at least 30 inches above the water line of the boiler, or more if possible. This was necessary so as to have a head or column of water sufficient to lift and keep open the check valve between the trap and the boiler during the discharge. The operation was simple enough. When the valve in the upper or receiving pipe from the pot below was turned on, and the air expelled from the sphere by means of an air valve placed on the top of the trap, the water from the heating coils was driven up by the steam pressure in the receiving pot, placed so as to drain all the return pipes in the building. After the sphere was filled it became heavier than the counterbalance weight on the other end of the frame, and went down. In doing so it tripped the equalizing valve and let live steam into the trap from the boiler, as before described. The pressures in the boiler and the sphere being balanced, the difference between the hight of water in the two gave the power necessary. The water from the trap lifted the check valve in the lower or discharge pipe and passed into the boiler. As soon as the sphere was relieved from the weight of water it became lighter; then the counterbalance went down. This action tripped and closed the steam valve on the top of trap. The steam within the sphere then condensed, and the sphere was ready to be charged again and repeat the work. The principles of action described herein cover the mode of working of all this class of traps. The difference is merely in the form and arrangement of parts of the mechanism.

Another form of trap was brought out a little later—called the bucket or stationary trap. There was a pot or water receiving vessel of greater hight than diameter, with an extension or ear on one side in which worked a lever. The pot was made in two parts and bolted together. Within this pot was an arm or lever with a square hole in one end through which passed a steel spindle. One end of this spindle rested and worked in a brass nut; the other end rested in and passed through a stuffing box to the outside; there it was fastened to a boss by a set screw. This boss formed a part of a lever with a long and a short arm. In the latter was a pin. Within the trap shell was an open bucket filled with pine wood, with a loosely fitting cover over it. This acted as a float When the water entered the trap a deflecting plate, which was hung beneath the water inlet, deflected the current to the sides of the shell. To the outer end of the steel spindle, after it passed through the stuffing box, was attached a lever. From one end of this lever an adjustable counterbalance weight hung. On the front of the pot was placed the equalizing steam valve case, made of brass and containing two small valves, which were tripped by arms through which a spindle passed with a square on it. A round part of this spindle rested in a bearing in the casing, and the other round end passed out through a stuffing box. To this a plain upright lever was fastened.

Above the valve box was pivoted a rocking arm, the ends of which were raised or lowered alternately. Four round weights, connected by two spindles and side rods, rested on this arm and rolled from one end of it to the other. As they moved from the boiler in passing along, the spindles struck the lever which was made fast to the equalizing valve spindle, and thus opened or closed the steam supply or exhaust valves alternately. The rocking arm received motion from the pin on the short lever which worked in the slot in the rocking arm. A small steam supply pipe connected with the valve box. An equalizing pipe from the valve box to the top of the trap case, and an exhaust pipe with a stop valve in it from the bottom of the valve case, completed the parts. A pipe of ample size connected the receiving pot, which was located below the returns on the floor, and the top of the trap, with a check valve at the highest point. Another pipe of larger size was connected with a tee beneath the bottom of the pot. A short pipe from this extended down, and a flange on the end of this formed the support of the trap. Another or discharge pipe, with a check valve in it, connected with the tee below the trap and ran into the boiler direct.

The trap was usually placed upon the boiler wall, or on some suitable support near by. The action of this trap was like the other to some extent. The water of condensation from the heating system, being collected in the lower receiving pot, was forced up into the trap shell above. As it entered and collected the bucket was lifted by it, aided by the counterbalance weight on the lever outside. The pin in the short arm of the same lever, working in the slot in the rocking arm, raised or lowered it, thus running the rollers from one end of it to the other, one or the other of the rolling spindles striking the lever on the equalizing valves and opening the steam or the exhaust. The action of the machine was noiseless, prompt and efficient, and was the type from which all the bucket machines now made have directly descended.

While the working of both forms of trap was good the Blessing trap required a good deal of space, and depended upon the surface condensation of the equalizing steam alone, and thus required more time to fill. The bucket form of trap occupied less space, and by promptly discharging the equalizing steam instead of waiting for it to condense quicker action resulted and much more work could be obtained from a trap of the same size and cost. There are many different kinds of the bucket form of trap now on the market.

The automatic return steam trap can best be applied to high pressure systems, or a combination of high pressure on the boilers and low or medium pressure on the coils. It is not desirable or necessary in ordinary low pressure work.

Q. 13. What is an automatic steam pump governor or regulator? How is it made, set in its proper position, and connected up for work? What are its special uses and how applied? How many different forms of it are now or have been made? What were or are now their special advantages? To what systems of steam heating is it usually applied?

A. An automatic pump governor or regulator, as now used, is a device for the regulating of or governing the speed of a steam pump by reducing or increasing the supply of steam to it. Its present application is particularly to the returning of the water of condensation from the pipes of a steam heating system direct, or through a receiving tank and a steam pump to the boiler.

Two styles are mainly used. The former represented by the Blessing, Kieley and others; the latter by the Worthington, Dean and many others. In the first form the governor and pump are usually separate machines, although set near each other; in the latter form the receiver, regulator and pump are all on the same bed and form one machine. In the Blessing and like systems the form and construction of the apparatus are closely that of their steam return traps, placed on or below the boiler floor instead of, as with the trap, on the wall above. The rise and fall of the bucket inside of the trap case through the lever and spindle outside, or directly, work the steam supply valve to the pump. The return water comes into the top of the trap shell and passes out to the pump through the pipe connected with the bottom. The relative amounts of the inflow and outflow regulate the action of the apparatus.

The Worthington and other like forms of apparatus for this purpose are a duplex steam pump, a receiving vessel with a float and balanced valve inside, connected with a steam supply pipe and valve to the pump, so arranged as to automatically operate the pump as the amount of condensation increases or diminishes. The forerunner of all these machines was a feed water regulator patented and introduced over 25 years ago by Robert Berryman, the inventor of the well-known feed water heater which bears his name. There was a hollow sphere, a counterbalance weight on the opposite end of a lever, two long pipes, one from the top, the other from the bottom of the sphere. The upper was connected with the boiler at the water line, the lower below it in the water. A lever was attached to the steam supply valve of the pump, which was operated by the rise and fall of the sphere through a connecting link. The apparatus was located on or near the boiler. When the water in the boiler was up to its proper level the mouth of the upper pipe was covered, the sphere was full of water and down, the steam valve was shut off and the pump stopped; as soon as the water fell below its proper level the mouth of the upper pipe to the machine was opened, the steam rushed into the sphere, the water in it returned to the boiler, the sphere rose up, the steam supply was opened, and the pump started to feed again. It can be readily seen how all forms of pump governors could follow from this.

Each maker of such an apparatus claims certain advantages, but simplicity, compactness, reliability in working and facility for prompt repair are essentials.

Q. 14. How many different methods have been used for returning the water of condensation from a steam heating system back to the boiler ? How many are now in use and how arranged ?

A. There have been four different methods used to return the water of condensation from high pressure systems of steam heating back to the boiler with the least possible loss of heat:

1. By direct return to the boiler, the full boiler pressure being carried from the boiler, through the mains, risers, heating pipes and returns back to the boiler, the same as in a low pressure gravity apparatus.

2. The boiler steam pressure, or less, is carried throughout the heating system and the water returned to a Nason common steam trap, from thence delivered to an open tank, a tank with only atmospheric pressure or but little more. From this it is pumped into the boiler.

3. The boiler steam pressure, or less, is carried throughout the heating system, and the water of condensation from the radiators and coils, both above and below the water line of the boiler, is collected in a receiving pot and by means of an automatic return steam trap is returned to the boiler.

4. The boiler steam pressure, or less, is carried throughout the heating system and the water of condensation from all sources is taken into an automatic steam pump governor or receiver, and from there returned automatically to the boiler by the steam pump.

All of these systems are now in use, but the tendency is, in the large steam plants, to carry high steam pressure on the boilers for power purposes, and moderate or low pressure steam in the heating pipes.

Q. 15. Is there any difference in the sizes and modes of running the steam and return mains for high pressure from those for low pressure? If so, why?

A. There is some difference between the sizes of pipe used for high pressure and low pressure. Steam of high pressure is of a higher temperature, consequently the same volume of steam will give out more heat under high pressure than low. So it has been proven in practice that the amount of radiating surface required to heat two rooms of the same size and exposure is about 20 per cent. less for high pressure than for low. High pressure varies from 50 to 100 pounds per square inch or more. The manner of running the pipes is about the same for high pressure as for the best class of low pressure work, proper inclination and a separate relief pipe from the end of the steam main being returned to the boiler direct, with separate globe and check valves, &c.

Q. 16. Is there any difference in the sizes, mode of running, or number of the steam risers and returns needed in high pressure from those in low pressure heating?

A. The number of risers and returns run for high pressure are the same as for low pressure two-pipe work. The sizes may be made less, but the best practice requires pipes of ample dimensions.

Q. 17. Is there any difference in the radiators or coils, their manner of setting up, arrangement or construction for high pressure and those for low pressure heating ?

A. There is no difference between the radiators and coils used for high pressure and those for low pressure. The construction, arrangement and connections also are the same.

Q. 18. Is there any difference in the valves used, their arrangement or connection with the radiators or coils ?

A. The valves used in high pressure steam work are usually made of heavier pattern than those for low, although the better class of contractors use the better grade of valves for both.

Q. 19. What is the difference in the expansion in the pipes when used with high pressure and when used with low pressure ? How should it be provided for ? What are the advantages or disadvantages of high pressure ?

A. The expansion of metals is dependent upon the temperature they are subjected to, and as high pressure steam conveys more heat than low pressure, it follows that the expansion of iron pipes containing the former must be greater than those containing the latter.

It should be provided for by what are called breaks in the mains, or change of direction, or any other simple means, but so-called "expansion joints" or sleeves are seldom effective and always troublesome. High pressure steam has its advantages in cases where the distance from the point of generation to the point of distribution is great, where the amount of radiating surface is necessarily limited, and where power is required.

CHAPTER X.

HIGH AND LOW PRESSURE STEAM HEATING AND POWER PLANT.

Question 1. In the past practice in steam heating how have the systems varied, and what is now considered the best practice for the best classes of buildings ?

Answer. In the past, and even at present, steam heating has been applied in a great many different forms by different designers and makers, but they may be generally classed under low, medium, and high pressure, exhaust and exhaust combined with live steam. All these grades of pressure have been used with direct, indirect, direct-indirect, or in some manner combined and modified.

In the beginning, low pressure steam was chiefly used in private dwellings and buildings of moderate sizes. In factories and large buildings generally either exhaust steam was utilized, or full boiler pressure, with a direct return to the boiler of the water of condensation, was the practice. This was then of moderate pressure, say from 40 to 60 pounds per square inch. As the pressure carried on the boilers for driving machinery, elevators and electric light plants, etc., for economical working was increased to a much higher point, it was considered unnecessary or inadvisable to carry full boiler pressure on the heating system; so reducing valves or steam pressure regulating valves were brought into use.

With these appliances and a large reduction of the working pressure in the steam pipes came the necessity for some means for returning the water of condensation back to the boiler. The receiving tank with a pump was among the earlier arrangements, but this was not automatic and gave trouble to regulate, and in many other respects was unsatisfactory. Then came the automatic return steam trap, having its own receiver with a closed circuit; then one with a broken circuit and exhaust quickening action was used. Next came the steam pump governor, being its own receiver and regulator combined. Finally the receiver, regulating valve and steam pump were combined, forming one compact machine. Each step has been made in compliance with the necessities of co-ordinate branch work, and they have become so intimately bonded together that it is very difficult to say where one begins and the other ends. It is therefore becoming more important each year that any one attempting to practice in

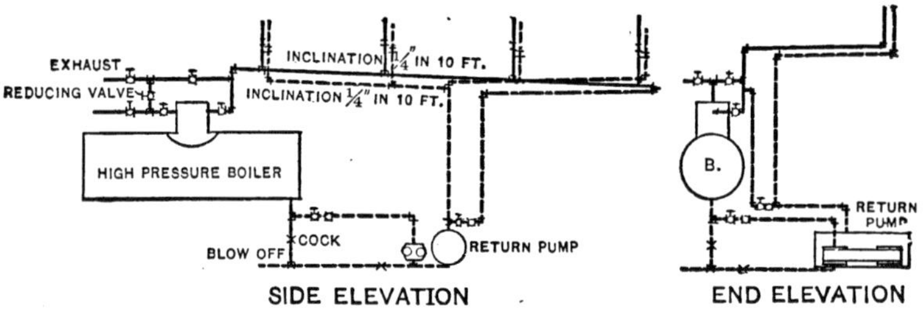

Fig. 81.—Two-Pipe Exhaust and Low Pressure System.

steam heating or practical steam engineering should be well trained in both branches of the business, for they must assume the responsibilities, at some time, of both installing and practically running the apparatus.

In the best classes of modern buildings the exhaust steam from the steam pumps for the elevators, the water supply pumps, and electric light and power engines are all utilized for heating the feed water and the buildings. Then, if this is insufficient, provision is made in the steam main, by means of a pressure regulating valve, by which any deficiency may be made up by an additional supply of steam taken directly from the boiler. Fig. 81 shows a two-pipe exhaust and low pressure steam heating system.

Q. 2. In designing and arranging a first-class modern steam heating and power plant what parts are required, how and where located and arranged?

A. In designing a modern first-class steam heating and power plant the parts and appliances necessary for a complete apparatus, as above described, are the high pressure boilers with all castings, set up in brick work, the smoke flues connected with the main smoke stack, fire tools, etc., complete; all necessary pipe connections around and between the boilers, as feed and blow-off pipes, steam heating and power mains and connections and valves, safety valves, damper regulators, steam and glass water gauges, feed water steam pumps and receiving tank, or receiver and pump combined, feed water heater, a blow off tank, feed water injector or inspirator, as it is sometimes called; oil, steam and water separator; all steam, exhaust and drip pipe connections around the steam pumps and engines; main exhaust pipe carried up to the roof, with an exhaust cap and sometimes back pressure valve. The drip pipes run to a tank cesspool or sewer outside of the main sewer trap to the building, the separator to be placed in the main power pipe to engines, etc., with all valves, cocks, check valves, etc., complete; the reducing or steam pressure regulating valve placed in the steam heating main, usually near the boiler; the steam supply and return mains; all branch pipes, risers and connections complete. Also the radiators, coils, radiator valves, air valves, air pipes, painting and felting. These constitute the main parts and appliances in a modern first-class steam heating and power apparatus.

Q. 3. What kind of boilers are used, and in what respects, if any, do they differ from those used for low pressure heating?

A. The boilers used are of the horizontal tubular type largely; sometimes the Babcock & Wilcox, of the water tube style, or the Sterling, Heine or other form—all built with extra care and of extra strength for high pressure work. They differ from those used for low pressure chiefly in the extra strength of parts.

Q. 4. Is the setting the same ? If not, in what respects does it differ ? How are the boiler smoke flues connected with the main smoke stack ?

A. The general arrangement of the setting is about the same as for low pressure with the same class of boiler. The smoke flue connection between the boiler and the main smoke stack is sometimes made by a brick arch over the top of the boiler, forming a flue connection with the main stack in the rear ; sometimes by a sheet iron pipe connected with the smoke box in front, and carried to the side or center and rear, wherever the main stack or flue is located.

Q. 5. What parts and castings are required for setting any of these boilers ? How made and set up in their proper places ?

A. The parts and castings are the same as for regular high pressure work, as before described, with ornamental cast iron sectional fronts, with smoke, fire, ash pit and flue cleaning doors ; flame and arch plates, grate bars and bar bearers, buck stays, tie rods, etc.

Q. 6. What attachments and regulating parts are required ? Where set up and how connected ? How are the dampers of special construction set up and applied ; their advantages ?

A. The attachments are the same as for a regular high pressure apparatus, as before described. The regulating attachments are of various kinds, formerly a steam damper regulator, after the style of the Clark, and very much like the one used for low pressure, only of stronger make, with heavier bowls, diaphragm, lever and weight, pivoted and set upon a stand and the end of the lever attached to a rod, which operates the pivoted damper in the boiler smoke flue. This regulator is a very simple and fairly effective device. Recently damper regulators have been made which are more sensitive, working within even ½ pound range of pressure per square inch on the boiler with a good draught, the old Clark being satisfactory with variations of from 2 to 5 pounds. But the simplicity of the former is lost in the latter, the pressure of steam from the boiler acting directly on a piston or indirectly through a column of water formed by the condensation of steam on the piston, or its equivalent, a diaphragm of metal, with a counterbalance, weight, &c. This causes the action required through the lever on the pivoted damper and increases or checks the draught.

Q. 7. What parts and appliances are generally used for feeding water to the boiler, and in what respect do they differ from any other, and why ?

A. The same appliances are used for feeding water to the boiler as in any high pressure work. Sometimes a steam pump alone, of Worthington or other good make, but often a feed water injector is added, in case of an accident to either. The steam pump for supplying water for use in the building is also so connected with its water supply and discharge pipes that it may be used for either purpose. In many cases a steam return trap has been used for that purpose, with or without a feed water heater. When an extra water supply to the boilers is required for various purposes, or if all of the steam is not condensed in the water returned to the boilers, a jet of cold water is admitted into the receiving pot and fed to the boiler with the return water from the house, even if a heater is not used, thus raising the temperature of the water admitted, but lowering that of the returns from the heating apparatus, and reducing it to a more moderate degree when it rises in the trap.

Q. 8. What is a "separator," how made, applied and connected up ? What are its advantages in a steam plant ?

A. A separator or eliminator is a device used for separating the steam when passing out of a boiler to an engine, pump, etc., from the entrained water which is often carried along with it in its rapid flow from the steam space of the boiler to the engine cylinder. A like device is often used for the purpose of freeing the exhaust steam from oil after it passes from the engine or pump cylinder toward the heating apparatus, so as to prevent it from collecting in the pipes, coils and radiators, and eventually clogging them up. The principles of action are the same in all these devices. The breaking up of the current of steam and changing its direction, sometimes giving it a spiral motion, are the most effective means used. This with the slight retardation which results from it has proved most successful. The forms used have been most varied—corrugated, spiral shaped surfaces and tubes, wire nettings, small tubes, cones with spiral wings on its surface, and many other forms ; all of these, with some form of water or grease receptacle placed below, fitted with a glass water gauge and a drip pipe, and connected to a common steam trap of some form with valves and pipe connections to the receiving tank or sewer. These have proven of great assistance to high speed engines, but a properly designed, constructed and managed boiler should not foam or carry over its water, but furnish dry steam. The forcing of boilers, too small to do the work, is more often the cause of trouble than anything else. So the separator, or eliminator, as it is sometimes called, acts for the engine as the steam safety valve does for the boiler, and prevents serious trouble by taking out the water before the steam enters the cylinder.

Q. 9. How are the steam pumps set up and connected with the boiler and water supply, and how best utilized ?

A. The steam pumps for various purposes are located where most convenient to the places and for the purposes for which they are to be used, and not too distant from the source of steam and water supply. They are usually set up on a solid brick foundation capped with a fine axed blue stone top from 3 to 5 inches thick or more, with holding down bolts and nuts. The pipe connections are, with the main steam pipe pump supply from the boiler, of a size larger than the opening into the pump ; also all drip pipes from the front and rear of each cylinder are united into one and carried to the sewer or cesspool.

The exhaust pipe connections are made first with the feed water heater, if there is one there thoroughly dripped, then run to the main steam heating pipe and connected with it beyond the reducing valve with a branch from the exhaust main. Both of these pipes have valves in them and are run to and connected with the main exhaust pipe rising to the roof at some convenient point, with a drip pipe and siphon trap at its bottom When the exhaust pipe rises above the roof an exhaust cap or pot is screwed on it and a relief pipe run from it to the house leader.

Q. 10. Are feed water heaters used ? How placed, connected up, and operated in connection with the steam feed pump to the boiler ?

A. Feed water heaters are invariably used and located as conveniently near the engines, pumps and boilers as possible. Sometimes they are set vertically on a concrete or brick foundation ; sometimes placed horizontally on a frame work near the floor, and again suspended from the floor beams above. In either position the heater is connected to the cold water feed supply from the pump, and the hot water discharge pipe from it is connected with the boiler with valves, etc., complete; the main exhaust pipe, a branch from it, with a valve on each at their junction, is carried directly to the heater and connected with it on the side or top, depending upon its construction and position. From the opposite side a similar pipe with a relief pipe for the water of condensation is connected with it. From thence it is run back to and joined with the exhaust main to the house, or to the roof, with valves so arranged that the exhaust steam may be carried directly from the engines and pumps to and through the heater and from thence to the heating main, to the roof, or it may be run directly from the engines and pumps to the heating main, or direct to the roof.

Q. 11. What is a steam pressure regulating valve? How utilized and connected up? How made, and of what value to a steam heating plant? How many different forms are there and their advantages?

A. A steam pressure regulating valve is a device used for reducing and regulating the pressure of steam in certain pipes, usually for steam heating, and is generally applied so as to make up any deficiency which may occur when utilizing the exhaust steam from the engines, pumps, etc., for heating. It is so constructed and arranged that the volume of steam required may be fully maintained, but the limit of pressure not increased. When the exhaust is not used it simply maintains a uniform pressure in the pipes with steam direct from the boiler by opening or closing the valve to suit the demand for steam.

Among the first of these was one put upon the market about ten years ago by Handren & Ripley, marine engineers, of New York City. To describe it briefly, it resembled a damper regulator turned upside down, with a valve box above it. The spindle passed through the lower part and rested on a rubber or metal diaphragm. As this spindle attached to a balanced valve was moved up and down by the action of the steam from the lower pressure side of the valve the controlling valve was opened or closed, more or less, thus increasing or diminishing the volume of steam which passed through it.

The construction of all those which have followed are on the same general principles—sometimes a rubber or corrugated metal diaphragm, inclosed in a case, one, or even two, spiral springs, and a balanced steam valve; in others a lever and counterbalance weight in place of springs, all so arranged that when set to carry a certain pressure in the mains or elsewhere as soon as the pressure rises above that point the valve closes down and diminishes the flow of steam. It is of great utility in exhaust steam heating, as it would be very annoying to be called constantly to watch and open or close the supplemental valve by hand. It is also used for many other purposes.

Q. 12. What is a hot water receiving and blow off tank? How made, where located, how connected with the return water mains and the steam pump, and of what special use and advantage is it as applied, and what is its general relation to the rest of the apparatus?

A. A hot water receiving and blow off tank is a receptacle into which flows the water of condensation from the pipes of a heating system, and is from there put back into the boiler by a pump or other means, as before described. It is usually made of sheets of wrought iron or steel riveted up, with heads of like material. All the return pipes are brought in at the bottom, or near it; the suction pipe from it to the pump is placed a little above the bottom, and the blow off to the sewer at the lowest point. The vapor or relief pipe, to prevent the accumulation of pressure, is connected at the top and carried to the roof, or into the rising exhaust pipe if not much used. The tank is sometimes set in a pit made below the floor level; at other times above the floor, on brick piers or an iron stand, as the location may require. In a low pressure apparatus, as this really is, the tank is very desirable, as it takes the place virtually of the boiler in the smaller system. Sometimes pressure is carried in the tank and a safety valve is placed in the relief pipe. Under such circumstances we have a complete low pressure apparatus from the reducing valve through the steam mains, risers, coils, radiators and return mains back to the tank, which represents the low pressure boiler. Valves and check valves should be placed in all pipes where desirable around the tank, so that it may be shut off completely when necessary.

When used as a blow off tank for the boilers, as required in New York City, the tank should be connected with the blow off pipe from the boiler by cocks, so that the water from the boiler may be blown directly into the tank and from there into the sewer, or to the sewer direct.

Q. 13. How many different steam heating and power supply mains are generally run out from the boiler, and for what purposes? What is the advantage of taking out more than one?

A. As generally arranged one steam heating and one power steam supply main are run out from opposite sides of the dome, usually front and rear, of each horizontal tubular boiler, or like connections in other forms of boilers. They are run in this way to prevent conflicting action, which would likely occur if a supply for two different purposes were taken from one pipe. The dome forms a large reservoir in direct connection with the boiler and furnishes full boiler pressure to either pipe; besides the rush of steam is not all to one side, but more evenly divided.

Q. 14. Is there any difference in the sizes, in designing or manner of running the mains for a low and high pressure combined system, or an exhaust steam supplemented by live steam from the boiler system of heating, from the regular practice when arranged for low pressure alone?

A. There is no difference in the sizes, in designing or in the manner of running the mains in a first-class low pressure system and a combination high and low pressure or exhaust system. All parts, except those specially required for reducing pressures, &c., are the same.

Q. 15. Is there any difference in the number, sizes, or

running of the risers and their returns and branch connections? If so, in what respects, and why?

A. There is no difference in the number, sizes, or the running of the risers, returns or branch connections between the combined high and low pressure and exhaust systems and the best low pressure work.

Q. 16. In very high buildings how is the expansion in the risers provided for, also in the lateral branches and connections?

A. In very high buildings the expansion in the rising and return pipes is provided for by one or more breaks, with swinging joints. These swing pipes are usually run between the beams and under the floor when possible, or in other suitable places above the floor. Other devices have been proposed and tried, such as corrugated disks of copper, &c., so placed and connected up with the rising pipes as to take up their expansion, but the breaks in the pipes, when properly put in and in sufficient numbers, accomplish the desired object in a simple manner.

Q. 17. In what respects, therefore, if any, does a modern high class system differ from the old low pressure system after the steam passes the reducing valve?

A. In reality there is no difference between the systems above described after the steam passes the reducing valve, as it is then, as usually operated, only a low or medium pressure system, and it should be so treated and managed.

Q. 18. Is the regular high pressure system, with direct return to the boiler, much used now? If not, why not?

A. The regular high pressure system, with a direct return to the boiler, is not much used now. The strain on the pipes is great, and there is no advantage in proportion to the disadvantage; so steam of lower pressure is carried in the pipes and some form of artificial return, such as has been herein described, is used, and better circulation and effect are thus obtained.

Q. 19. What kind of radiators are usually used in modern first-class buildings? In factories or other work buildings?

A. In modern first-class buildings, either the wrought iron vertical pipe radiators or cast iron vertical pipe or section radiators, which are of many shapes and designs. In factories, stores, work buildings, coils of all kinds are used, some placed along the walls, others suspended from the ceiling.

Q. 20. What class of valves? How made and finished in each case?

A. The valves are generally some form of a soft disk valve. They are generally nickel plated, with polished trimmings, but the plain metal valves are used also.

Q. 21. What air valves, with or without air pipes, and their connnections are used? How made?

A. Air valves are generally used, some depending upon the expansion of metals, others of some composition. Some forms are without air pipes, but the safest and best arrangement is with them. These are connected into mains in the cellar and carried to a sink or cesspool

Q. 22. How are the radiators and exposed pipes above cellar finished? The iron work and pipes in the cellars?

A. All pipes and iron work below the first floor are usually black varnished. On the first, and above, gold or silver bronzed, except in factories, where in work rooms they are black varnished.

Q. 23. Is it advantageous to have pipes in the cellar felted or covered? If so, why? What means are used, different kinds, and their advantages?

A. It is advantageous to have the mains and other pipes in the cellar covered, when the heat is not required there, and in certain cases it is even preferable to felt the pipes, and put coils, many different kinds are used, in for any extra heat required.

Q 24. What other appliances are used in these modern buildings? Is the steam contractor usually required to make the steam exhaust and drip connections to any of them?

A. In later years the introduction of elevators oper ated by steam, hydraulic power or electricity has added to the responsibilities of the steam heating contractor, as the steam supply, exhaust and drip pipes, with their varied combinations and complications, call for the exercise of much ingenuity, skill and good judgment in their arrangement, and the tendency is to increase rather than diminish this responsibility. Contracts for all the pipe connections for the steam, drip and exhaust to elevator pumps are usually given to them.

Q. 25. Are the elevators and their connections ever included in the heating and power contract?

A The contractors for steam heating are seldom called upon to take the contracts for elevators, but the tendency is to concentrate as much as possible all contracts under one head. The results, however, are not promising, as the general contractor wants a profit for his risk. The sub-contractor must either charge so much more or slight his work, for competition is too close to admit of much margin for profit.

Q. 26. Are the electric light and power engines and their connections ever included in the steam contractor's bid?

A. The electric light and power engines are sometimes included in the steam heating contract; the boilers are almost invariably placed with them; the steam supply, exhaust and drip connections are usually placed in their hands.

CHAPTER XI.

EXHAUST STEAM HEATING.

Question 1. How is the exhaust steam from the elevator steam engines or pumps and the various boiler feed and other steam pumps and electric light engines in modern buildings utilized ?

Answer. Exhaust steam from the engines of a power plant has long since been used for heating mills and factories of all kinds where the power used was constant, but its introduction into the heating system of stores and other public buildings, alone or in combination with live steam direct from the boiler, is of more recent date. The introduction of steam pumps and power engines has furnished the amount of steam required, and economy in working demanded some further service for it after it passed through the engines, &c., as the amount of heat it still retained was large and valuable. Modern and improved modes of utilizing it to the greatest advantage are so complete that in many cases little or no aid is required direct from the boiler.

Q. 2. What is the benefit in its use alone or with live steam direct from the boiler ?

A. The benefit derived from its use is simply the difference between utilizing the exhaust and throwing away into the atmosphere an amount of heat often sufficient to warm the whole house and then abstracting that amount of heat directly from the boiler in the form of live steam to make good that waste. It would be about as rational as to throw into the ash pile a shovelful of coal for every one placed in the boiler furnace. Of course where there is not a sufficient amount of exhaust steam for the purpose required it must be made up from some source, and the mode and manner described before has proven most satisfactory.

Q. 3. How is it best applied, alone or in combination with live steam?

A. As to which is the best plan, to use exhaust steam alone or in conjunction with live steam, undoubtedly if you have a sufficient amount of exhaust steam there is no necessity for aid, but there are days and hours in extreme weather when the demand may be increased and the supply diminished. Then the supplementary supply is most convenient and even absolutely necessary.

Q. 4. How many different ways is it generally utilized? How constructed and arranged?

A. There are two ways of utilizing the exhaust steam: 1. When it is introduced into the regular steam heating system as usually set up, from below or through the steam mains and risers. 2. When the main exhaust pipe from the engines and pumps to the roof is utilized by placing a back pressure valve near the top and taking off the distributing supply mains from it by outlets placed near the ceiling on each floor, then running around and dropping down to each coil (Fig. 82). The returns from each line of coils are collected into one return pipe line and carried down to the cellar, there united with the main return and carried to the receiving tank, the water of condensation from there being returned to the boiler.

Q. 5. What new arrangement has been made for facilitating the circulation of exhaust steam in the heating pipes? How is it arranged and operated?

A. A new arrangement and mode of operating all steam systems is by exhausting the air from the radiators, coils and pipes, thus removing one of the chief causes of sluggish circulation and obviating the necessity of carrying steam pressure above the atmosphere in the system; in fact, many plants are now being successfully worked below atmospheric pressure. The removal of the air from the system enables the steam to more fully reach all parts of the radiating surface, consequently the transmission of heat is more rapid. When working in the ordinary way the pressure of steam within the coils, radiators and pipes forces or pushes the air to the point where the air valve is supposed to take it off. But air being elastic and expansive, very often refuses to be so handled; so it is like trying to push a rope before one to a point where it is

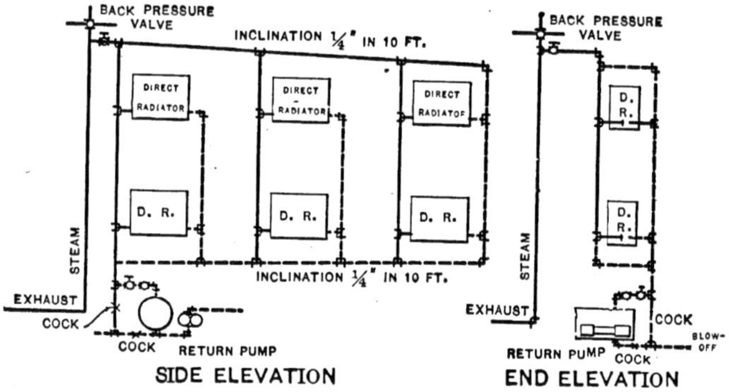

Fig. 82.—Exhaust System —Main on Top Story.

needed. With the new system it acts by induction, so instead of a push it is a pull, and is so much more effective.

The parts used and their application are most simple. An ordinary ejector (not injector) of small size is connected with the end of the air main in the cellar. Into this main all the air risers and branches to the automatic air valves in the system are connected. To this ejector is connected a small steam pipe, taken from any proper source, or a small water pipe from a high pressure hydraulic elevator tank or other good source of water supply under pressure. This being introduced into the ejector and passing out, induces a current and a partial vacuum in its wake, and so draws the air out from the pipes and radiators with it. After the radiator is exhausted of air the steam rushes in and the automatic air valves close. If it should accumulate again, the same action is repeated.

CHAPTER XII.

POWER FAN OR BLOWER SYSTEM OF STEAM HEATING AND VENTILATING.

Question 1. What is the power fan or blower system in steam heating and ventilating work? How arranged? What parts are required and how made? Has it been much used and how long?

Answer. The power fan or blower system in steam heating and ventilating work is one of the best in the whole range of heating and ventilating appliances. Its object is the furnishing of fresh and pure warmed air to the various rooms and apartments in buildings, generally of a public character, and of large size. The high first cost and the after care and attention required prevents its introduction into smaller houses. The parts required for a complete apparatus are the cold fresh air supply ducts, the warm air distributing chambers and ducts, and outlets or delivery controlling valves or registers, as they are called, the heating coils, a steam engine, or other form of propelling power, a power fan, and all parts and attachments necessary to properly operate the same.

When a means of cooling the incoming air in summer is required, instead of heating it, as in winter, an ice rack and chamber are constructed, or some other artificial means of lowering its temperature is devised.

Q. 2. How many different forms or arrangements of parts of this sytem have been used? What advantages have each?

A. There have been about four different modes of general arrangement of this heating system used, besides modifications of the kinds, class, and the materials used for radiating surface.

The first is where the cold air is drawn directly into the power fan from the cold air ducts, and by the fan is driven through the heating coils, etc., located near it, the heated air then passing into the main pipes or chambers for distribution and through the smaller pipes throughout the building.

The second is where the cold air is drawn from the cold air ducts through a primary heating coil and is slightly warmed before passing through the fan, and by it forced into the main heating coils to be fully heated, and from thence passed into the main pipes or chambers for distribution.

The third is where the cold air, from the cold air ducts, is first drawn by the fan through the heating coils, etc., located near it, then forced by it into the chambers and the large and small pipes for distribution.

The fourth is where the cold air is drawn by the fan from the cold air ducts direct and forced along other large and small cold air ducts to the point where the radiating surface is located, which is usually at the foot of the hot air rising pipes; the cold air is then passed through each coil or heating stack, is heated, and rises up to the point of delivery in each room.

The advantages claimed for the first arrangement are that the air being cold when it enters the fan a greater volume of it can be forced through it in a given time than if the air is heated and expanded before it enters the fan. The advantages claimed for the second are that the cold air, before it reaches the fan, is drawn through the primary coil, or stack, in which any exhaust steam or other source of heat, which might otherwise be lost, is utilized, and the chill taken off the air, in very cold weather. This prevents the main coils or stacks from being frozen up In very moderate weather the primary coil would give heat enough to the incoming air without the steam being turned on the main coils. The advantages claimed for the third arrangement are that the cold air is drawn by the action of the fan through the main heating coils or stacks and thoroughly heated before it enters the fan, and then it is forced into the distributing pipes. This drawing action, it is contended, gives better results, even if the air has been already expanded, than the forcing of cold air through the heating coils.

Q. 3. How are the heating coils arranged or placed? How many different kinds are used and how made?

A. The location of the coils in relation to the air supply, power and mode of distribution are given above. The mode of construction and manner of setting up are varied. In times past the heating coils for this purpose were generally constructed of 1-inch or $1\frac{1}{4}$-inch wrought iron pipe made up in the form of box coils and set up in brick chambers at the foot of each rising hot air main, or in one large brick chamber near the fan. In latter days the brick chamber has been succeeded by the cheaper form of galvanized sheet iron, protected by wood or felt or some non-conducting material. The distributing pipes are almost always made of galvanized sheet iron, suspended from the floor beams above or placed on special supports.

The heating stacks are now often made up of a number of cast iron indirect heating sections, united by nipples or other modes of connection. In these can be used live steam from the boiler, direct, high or low pressure; exhaust steam alone or in combination; all completely under hand control by valves, etc., or arranged for automatic action.

Q. 4. How are the cold air supply inlets arranged? In what relation to the power fan and why so placed, and of what made?

A. The cold air supply inlets are arranged so as to take the supply from the point where the purest air can be obtained, from the roof of the building, or at an intermediate point, which is generally best, for if near the earth or from the roof, dust, foul gases from sewer pipes or other sources can be drawn in, but between these points the best quality of air seems to be obtained. These ducts are often built of brick as a part of the structure or separately, as may seem best. Sometimes they are made of galvanized iron, boxed in with wood or fire proofing. They vary in form with the circumstances and surroundings. As the power fan can draw air from any quarter its location, as regards the points of the compass, is not essential for success. As a general thing, however, the distance from the mouth of the opening of the fresh air duct to the point of its delivery to the fan is fully considered, as the greater the length of the duct the greater is the friction to be overcome, the power required, and the greater the first cost as well as cost of operating.

Q. 5. How are the hot air ducts arranged? Of what material made? How run and what special care is required in construction?

A. The hot air ducts are sometimes made of brick plastered on the interior or even lined with metal so as to make a smooth surface and lessen the friction of the air passing through them. Again, vitrified or glazed pipe has been used to advantage, as it has a smooth interior and is a moderately fair non-conductor of heat. In modern structures galvanized sheet iron pipes are used, and are run near the ceiling in the cellar or basement, suspended from the beams. They are of ample size in good work. All changes of direction are made without short bends, and the Y principle is universally used. The interior should be free from obstructions of any kind, including sudden diminishing of the diameters of pipe. Where required it should be by gradual inclines.

Q 6. How is the hot air carried to the different rooms and compartments to be heated? What relation is there between ventilation and heating?

A The hot air is distributed to the various rooms, halls and compartments requiring heat by vertical, horizontal, or slightly inclined pipes, and at the various outlets controlled by hot air registers. These pipes should be of ample size, run as straight as possible, with easy curves and smooth interior surfaces. In ordinary indirect heating, when there is an inlet for fresh, warm air, there should be an outlet specially provided for the outlet of impure or cooler air. With the use of the power fan, however, the means is given in a greater degree to force the circulation, so that if any reasonably fair chance is provided for exit we have a sure means of obtaining both heat and pure air. Thus, a supply of warm air and pure air by this means is made as positive and ample as may be desired.

Q. 7. How are power fans or blowers usually made and set up?

A Power fans or blowers are usually, under present designs, made of steel—the spindle, the arms, and the paddles or blades form the moving frame work of the fan or blower proper. This is sometimes set up and partly inclosed by a brick setting At other times it is inclosed in a sheet steel casing. The air enters at the center of the fan and the discharge is at the periphery, above or below, as may be desired. They are arranged with the air entrance on one side of the casing and the discharge at the periphery. When these fans are designed for exhausters their construction is somewhat modified.

Q. 8. By what means are they propelled? How are the steam engines, etc., made and set up, and how connected with the power fan or blower?

A. These power fans for heating and ventilating systems are usually driven by belts from shafting propelled by steam engines, water power, electric power or any other means most convenient. Sometimes the steam engines or electric motors are connected directly to the shaft or spindle of the fan. Some of the steam engines are of one or two cylinders, some are set horizontally, on a level with the spindle of the fan, others are set vertically and below the spindle, and others vertically and above the spindle. All of these are directly connected with the shaft or spindle of the fan. Others are driven by belting from a pulley on a countershaft, or directly from the fly wheel pulley of the engine.

Q. 9. What is the plenum system? How arranged for heating public halls, theaters, prisons, hospitals, etc.? Are these worked under pressure? If so, what are the advantages?

A. The plenum system is an arrangement of parts wherein the hot air from the power fan is driven into a close chamber, under a slight pressure, and from this the distributing pipes carry the air into the various parts of the building requiring heat, their exits being controlled by registers or valves. In theaters the chamber is generally beneath the auditorium floor, in churches in a similar position, and in other buildings wherever the gen eral construction of the building will allow. With this system the whole house is generally under a slight pressure. This being the case, the tendency is always for the air within it to flow outward, so that there can be no inward draughts, but all outward. This action is sometimes quickened by exhaust fans placed above the ceiling, between it and the roof. By increasing the pressure in the plenum, however, an ample outflow of impure air, with a proper number of outlets, can be insured, and perfect circulation secured at less expense. Wherever heat ing and ventilating are combined there must be some loss of heat in securing thorough change of air, but this is the price which must be paid for the better results

Q. 10. What other arrangements are sometimes made for combined heating and ventilating? Is the exhausting sytem ever used alone, or in combination with the force system?

A. Many different arrangements and forms of the parts and mode of distributing are made, and each claims some advantage. In one it is to force the hot air from the fan directly into the main distributing pipes, delivering it at or near the ceiling; in others about 6 to 8 feet above the floor. Again, outlets near to or in the floor are connected with the main exhaust air ducts, which lead to and are connected with an exhaust fan. Some ambitious individuals have introduced the heated air into theaters high up, even above the proscenium arch, and endeavored to draw it down through openings in or near the floor by an exhaust fan, thus attempting by force to reverse the natural order of flow of hot air. They must surely pay the penalty, either by failure or by the use of greater force to overcome the natural tendency of heated air to rise. It is difficult for some to see where the advantage comes in. The first claim is that in the working of this system a more even distribution of heat is secured without drafts. This is not reasonable, and many facts go to prove the contrary. "Keep the feet warm and the head cool," is an old saying. Because of the location of the exits in this system, the coolest air and the strongest draughts must be and are near the floor; the head would most likely be kept hot and the feet cold, which is about the case when in operation. The second idea is also fallacious, that the foul air in a room or apartment, being heavier and cooler, falls to the lowest point and should be at that point removed. This may be to some extent true when all the air in the room is cool, but heated air will rise to the highest point it can go, whether pure or impure. So when drawing it down by superior force it must become even more expanded and attenuated and likely to contain less of the life giving principle.

There are various modes of distributing the heated air, whether worked on the plenum or direct delivery system, with or without the aid of the exhaust system. In theaters it is sometimes admitted through the auditorium floor, by means of pipes with curved outlets having perforated tin coverings; also in the risers between the floors on which the seats in the dress and family circles are located; again at the side of the house and through registers in the aisles—in fact, almost every way one can imagine, as circumstances may require or fancy dictate.

The exhaust system is sometimes used alone. The fan is located either in the attic or in the cellar, depending upon whether the pull up or pull down system is used. In either case the air is induced to pass through the heating coils, becomes rarefied and lighter, and is further attenuated by the action of the exhausting fan.

Q. 11. What are the different modes of distributing the heated air in these buildings?

A. The different modes of distributing the heated air in buildings have been incidentally described. As before said, it varies in each system, and often in the same system, according to the ideas of those in charge of the work or the whims of owners, designers and constructors; but one general rule should govern all—seek to be natural or follow Nature's laws and you will be as near right as it is possible to be.

Q. 12. Is a complete system of heating and ventilating, in some form, much used at present?

A. A complete system of heating and ventilating combined has been often attempted. Some have been better than others, but there is still room for improvement. To better it is what all should strive after now. The first cost is the chief objection, but in time that will be reduced.

CHAPTER XIII.

DESIGN. ESTIMATE. SPECIFICATION.

Question 1. What are the first points to be ascertained and decided upon beforehand in "laying out," as it is often called, or designing a steam or hot water heating plant? Is the same routine required in every case, large or small?

Answer. The first point to be ascertained and decided upon before "laying out" a steam or hot water system is its geographical position, as this determines the average and lowest temperature during the winter season. The next is its local position. Is the building isolated, on a hill or in a valley, in a block, or on high or low ground? How does it face, north, south, east or west? Of what material is it to be or is built, of stone, brick, iron or wood, or a combination? What is the thickness of walls, and are they hollow or solid? If a wooden house, is non-conducting material used in its construction? Is there to be a combination of brick, stone and wooden frame work, and is the building to be one, two or twenty stories high, and for what purposes is it to be used?

Next obtain a drawing of the building showing the sizes of the rooms, their relation to each other, the halls, entrances, &c., and the hight of the various ceilings, window sills, &c. If there is no drawing in existence then make a sketch of the building, taking accurately the dimensions as called for, and lay out a plan yourself. Guess at nothing; time and money will thus be saved in the end. Then arrange the dimensions for each room, corridor and all other places to be heated separately, on a sheet of legal cap paper, as it is the best suited for the purpose, and figure out the cubic contents of each separately, so that you may be able to apportion each. No lump guesses should ever be resorted to by any one with any knowledge of the business. When this is done ascertain how much glass and exposed wall surface there is to the room, and calculate its cubic contents.

After this proceed to lay out the work on the plans, showing where the radiators for the direct or the direct-indirect heating, or the stacks for indirect in the cellar and the registers upstairs, are to be located; where the boiler is to be placed in the cellar and the direction and manner of running the mains, with all other statements as to sizes, &c., requisite. The same routine should always be gone through with in every piece of work, as it often prevents serious mistakes and is the very best record which can be kept of the work as executed. It saves a good deal of time and is a ready reference to experience gained, good or bad, successes or mistakes.

Q. 2. What does the ratio between the cubic contents of a room or compartment and the amount of heating surface mean? How is it determined and computed? How are the cubic contents of a room calculated? What is necessary for a clear and accurate determination of the proportion of parts required.

A. The ratio used in estimating the amount of heating surface required for warming any room or apartment is the relative proportion between the radiating surface and the cubic contents, and depends upon the kind of surface and surrounding influences, the location and size of the room or rooms, and the exposure. Many variable circumstances and conditions must be considered. A small room requires proportionately more heating surface than a larger one, as it has more wall or cooling surface in proportion to its cubic contents and the doors and window surfaces are generally proportionately greater, and so on. The cubic contents of an empty room are its length multiplied by its breadth and by its hight. There may be much or little furniture placed in the room, all of which affects the results obtained. Generally a liberal allowance is made for all contingencies.

Careful consideration and good judgment, backed by experience, are the best guides, as all data obtained from other sources are only approximate and a mere starting point. What would answer well in one case might fail in another, as modifying circumstances might arise which could not have been anticipated and given due care and attention. The amount of wall surface and the glass surface should be carefully taken into consideration.

Q. 3. What is the next step to be taken after this is done? Where should the heating surface be located for the best results for direct, indirect, direct-indirect?

A. The next step to be taken after the cubic contents of the rooms, the glass surface and the wall surface are figured up and placed in proper columns, is to find the best or most available location for the radiators. The heating surface should be shown on the plans, whether direct, indirect or direct-indirect, drawn to a scale and marked with the amount of square feet of surface, or the kind and number of pipes or sections. This saves much trouble in the end, and must tally with the specification. As to the best location for the heating surface, only one rule governs—place it as near as possible to the point where the least heat is likely to be or the cold comes from. Usually this is the north or west side, near glass or thin walls or doors opening outward; or on the other hand, where it may be allowed by the owner, architect, or future mistress of the house, whose wishes are often the closing argument.

Q 4. What proportions of parts are recognized as reliable and how should they be modified to suit variations in climate, surroundings and other circumstances for hot water, low pressure steam or high pressure steam?

A. The amount of heating surface in a room varies,

as aforesaid, with the size, exposure, amount of glass and wall surface, &c., also whether it is to be heated by low pressure steam, high pressure steam or hot water. The differences between these are in the amount of heat each contains, or carries, and can radiate within a given time, and this depends largely on the rapidity of circulation of the steam or hot water within the pipes, the difference in their temperatures and that of the objects and air surrounding the heating surfaces, and whether direct or indirect.

As there must be some starting point, low pressure steam with direct radiators can be taken as a standard to be installed in a fairly well built brick house to be used as a private dwelling, situated in a small town in a moderately hilly or rolling country, within, say, 50 miles of New York City, the house being isolated and facing north.

The parlor is in front, the dining room in the rear of it, the reception room on the opposite side of the hall to the parlor and the library back of it. Next to the dining room, as an extension, is the butler's pantry, with the kitchen beyond it. On the second floor four bedrooms are located over the main lower rooms, the bathroom and children's playroom being over the kitchen in the extension. In the attic are three sleeping rooms and one storeroom.

In this arrangement we have a type of a modern house of medium size and cost often found, which would be expensive enough to admit of the use of a steam heating apparatus. If there are bay windows in the sides of the parlor and sitting rooms, both of them facing north, with regular windows in front, the proportions of heating surface to the cubic contents would be about 1 to 60, properly distributed. In the dining room, on the west side, 1 to 70 ; in the library, on the east side, 1 to 65, or 70 as it often requires more heat in proportion than in the others. A parlor should be warm, a library likewise, a dining room a little less so, and a sitting room the same as the parlor. The lower floor of a house, including the halls, should be well heated, as the heated air rises from them to the upper floors. The bedrooms in this house could be put at about 1 to 75 or 80, and bathroom 1 to 60 or 70. Those on the west side of the house should be a little warmer, so in case of sickness they might be used. Some prefer cool bedrooms.

From this scale we may proportion others, as, for example, in a block the front rooms could be reduced five to ten points, the others in proportion to exposure. The radiators may be divided and with valves arranged so as to be able to use a part or all. For indirect heating all of these proportions should be increased from 20 to 30 per cent. or more for low pressure steam.

For hot water direct radiation about the same ratio is used as for indirect low pressure steam, and for indirect hot water as much as 20 to 40 per cent. above direct hot water. For high pressure steam about 20 to 30 per cent. less surface is used than for low pressure steam direct, and for indirect the same as for direct low pressure. As before stated, this is based upon the use of vertical tube radiators for direct and cast iron sections, prime surface, for indirect. When wrought iron horizontal coils are used a further reduction may be made for direct heating.

For hot water heating in a climate like that near New York, where it rarely goes below zero of the Fahrenheit scale, some designers use for parlors, sitting rooms and general living rooms of a residence 1 square foot of direct radiating surface for every 30 to 40 cubic feet of space, 1 square foot of surface for every 3 square feet of glass exposed, and 1 square foot of surface for every 30 square feet of exposed wall surface.

For library and dining rooms............ 1 to 30 to 45 cubic feet
Glass.................................... 1 to 3 square feet
Exposed wall................................ 1 to 30 square feet
Reception halls and rooms................ 1 to 40 to 50 cubic feet

Glass	1 to 3 square feet
Exposed wall	1 to 30 square feet
Main halls, first floor	1 to 30 to 35 cubic feet
Glass	1 to 3 square feet
Exposed wall	1 to 30 square feet
Chambers	1 to 50 to 65 cubic feet
Glass	1 to 3 square feet
Exposed wall	1 to 30 square feet
Nurseries	1 to 45 to 55 cubic feet
Glass	1 to 3 square feet
Exposed wall	1 to 30 square feet
Bathrooms	1 to 30 to 40 cubic feet
Glass	1 to 3 square feet
Exposed wall	1 to 30 square feet

The above is for bedrooms about 65°, library, &c., say 70°, and bath 75°.

School rooms	1 to 60 to 85 cubic feet
Glass	1 to 3 square feet
Exposed wall	1 to 30 square feet
Factories and stores	1 to 65 to 90 cubic feet
Glass	1 to 3 square feet
Exposed wall	1 to 30 square feet
Halls and churches	1 to 90 to 150 cubic feet
Glass	1 to 3 square feet
Exposed wall	1 to 30 square feet

For indirect coils and stack, &c., from 40 to 70 per cent. more than for direct radiation; some even carry it to 80 per cent. when cast iron sections are used. All of these, it must be remembered, are for hot water heating.

As to the sizes of the cold air ducts for indirect work, some designers allow for the first floor ⅝ square inch, net internal area, for every square foot of radiating surface in the indirect heating stacks ; for the second floor, ½ square inch for every square foot, &c. Hot air fines for the first floor to be ¾ to 1½ square inches net area for each square foot of surface in the heating stack ; for the second floor, ⅝ to ¾ square inch to every square foot. In hot air registers only the net area of openings should be allowed, which is about one-half to two-thirds of the nominal opening. The friction of air currents must also be taken into account.

Q. 5. After the amount of heating surface is determined, what is the next step? After the total amount of surface is added up, how are the size and proportions of parts of the boiler determined? What is the relative proportion of boiler heating surface for high and low pressure steam and hot water, and of the grate surface of the boiler to the absorbing surface?

A. After the amount of heating surface is provided for, the next step is to ascertain the size of the boiler required to supply it easily and promptly with the steam or hot water necessary. After the total amount of heating surface needed in square feet is added up, then the proportion of the heat absorbing surface in boiler to radiating surface in rooms, &c., is determined. These proportions vary with the evaporative efficiency of the boiler. All sorts of claims are made for the results from boilers of different makers. Some tests are made when the boilers are new and all the absorbing surfaces are clean; others carry water over, or foam, when hard driven; others are so much subdivided in their water space as to have but little solid water in them, nothing but spray and foam, and very little account is taken of the fuel consumed. It is merely claimed that they will supply so much radiating surface.

A good boiler for heating purposes should have ample grate surface and fire box, an extra amount of heat absorbing surface and a good draft. This means a proper size of smoke flue, and sufficient tube or smoke area for the hot gases to pass up easily at a moderate temperature and velocity from the grate. The heat should be thoroughly absorbed in passing through and out of the boiler, which is shown by a moderate temperature in the smoke pipe, say from 300 degrees to 400 degrees, or less if possible, while the fire in the meantime is bright and lively. The relative proportions of the absorbing surface, sometimes called "heating surface," in the boiler (if vertical or horizontal tubular styles are used, which are usually considered the standard), and the radiating surface in the building, including mains, branches, &c., are from 1 square foot in boiler to 4½ square feet or 5½ square feet, and even up to 10 square feet, of radiating surface in rooms. The latter some claim for their boilers. The proportion of grate surface to the absorbing surface of a steam boiler should be from 1 square foot of grate to 30 or even up to 40 square feet of boiler absorbing surface. With hot water the proportion of boiler absorbing surface to radiating surface in rooms is about 1 square foot to 8 or 10 square feet or even more.

Q. 6. The size of the boiler settled, what is the next step? What proportions are best between the main supply and return pipes and the radiating surfaces for direct steam heating, also for the risers and the connections between the mains and radiators? What proportions in indirect steam or hot water?

A. After the proportions of the boiler are determined, the next things to be considered are the sizes of the main supply and return pipes. A rule which some follow is to make the size of the low pressure steam main at the boiler from 0.78 square inch to 1 square inch of area for every 100 square feet of surface in the heating system and to run the steam with as few reductions as may be possible, and use eccentric reducers as far as practicable. The proportions should be carried out very full with due regard to the amount of surface called for at the furthest point, so as to maintain, as near as possible, the full boiler pressure at the end. The main return should start about one size less than the steam at the extreme point and be gradually increased in size to the boiler. In small systems the return is one size smaller than the steam main near the boiler, and in larger ones two sizes, or even less, at the same point. Under some circumstances it is best to run two or more separate returns, each with its own globe or check valves, and unite them near the boiler or enter it separately.

The sizes of the mains for hot water should be larger than for low pressure steam, and both the main supply and the returns should be of the same size. The reductions in the branches should be less, and both the branches and risers should be large in proportion. In the smaller pipes, for any system, ample allowances must be made for friction. For low pressure steam some designers use certain proportions of sectional areas of pipes for every square foot of heating surface they supply, but no risers should be less than 1-inch or 1¼-inch. For the riser, 0.006 square inch to 1 square foot; for the branches from the riser to main, 0.009 square inch; for the branches between the risers and radiators, 0.012 square inch. For connecting pipes between the steam supply pipes to the coils, when used with a fan blower, there should be 0.021 square inch for each lineal foot of 1-inch pipe or equivalent, assuming that the amount of condensation is fully three times as much in the fan blower system as it is in the ordinary indirect natural draft system. For hot water an increase in this is required.

For steam:

30 square feet in radiator, a.................. 1-inch steam supply
50 square feet in radiator, a.................. 1¼-inch steam supply
80 square feet in radiator, a.................. 1½-inch steam supply
Returns, one size less.

Hot water and single pipe connections one size larger. It is impossible to give exact figures, as judgment and experience must direct; these are only starting points.

CHAPTER XIV.

DWELLING HOUSE HEATING.

So much has been done within a few years to popularize the use of steam, hot water, and a combination of both, for heating and ventilating private dwellings, and so many of those who conduct business in small towns or villages are called upon to repair or install such work, that it is necessary that they should be well instructed in the principles involved and in the practice of erecting and operating this sort of apparatus. When properly constructed and erected there ought not to be the least trouble in the working. Unfortunately, however, there are many who have not had the advantages of a proper training in such work and must buy their experience at their own and others' expense, which often proves very dear to both. The selection of the style of apparatus, whether direct, indirect, direct-indirect, or a combination of two or all of them, is of the highest importance and should be intelligently considered, for it depends upon the construction, form and style of the house, its uses and general plan, the materials of which it is built, its location, surroundings, and the wishes and peculiarities of its future owners and occupants. True, it may be said, on general principles, if it is judiciously designed and works well it ought to suit every one, but this in practice is not so, and most architects and builders would rather design and erect several buildings to be used for business purposes than one private residence. In the one case men alone are to be dealt with; in the other, women, and all their fancies must be met and satisfied.

The first object in designing a heating apparatus for a dwelling is ample heat at all times and under all circumstances, all parts of the house to be of even temperature, without drafts (which are caused by differences in temperature), and a constant and regular change of the air in the various rooms. The amount of air sent to each room ought to be controllable in quantity and temperature, as living rooms and those generally occupied by the family and guests sometimes require more fresh air and less heat, or the reverse, than those used for sleeping purposes or which are only occupied by a few. Where the family is large or much given to entertaining the regulation of the amount of heat and air supplied is very important. Some will say, "This extreme nicety of adjustment is seldom necessary; just put in an abundance of everything and all will go right," but the day for such slipshod work is rapidly passing away.

Engineering skill should aim to furnish enough, neither too much nor too little, efficiency with economy. Owners will soon learn that intelligent competition in this line is as important as in architectural designs. He who gives most for the money spent is the right man to employ, or he who gives what is necessary or desirable for the least money. In order to properly design such an apparatus a man should be thoroughly trained in the work, both practically and theoretically; in fact, should, if possible, have lived in a house so arranged, and then he will know where the weak points are. If the comfort provided in their homes is one of the best evidences of the highest order of civilization among people, then this country should rank among the highest, if not in itself the highest on earth, for in no other country are the greatest of necessities, heat, light or even ventilation, better provided for than here.

Among the first modes of utilizing heat in a dwelling house was a brazier or open furnace in the center of the house, next an open fire place and chimney, then stoves, then hot air furnaces in the cellar, then hot water in pipes, then low pressure steam in pipes and radiators, direct and indirect, and now by a combination of steam and hot water with a hot air furnace, or modifications of all these. In all, the object has been to utilize the heat generated by the combustion of different kinds of fuel—charcoal, wood, bituminous or anthracite coal, and now we have come to gas, natural or artificial. The most difficult point to attain by many of them has been to advantageously apply the heat where it would do the most good. When limited to a single center of heat it radiated from that point; with the open fire place the number of points was then increased. The trouble with the hot air furnace was, and is often now, the excessive dryness and overheating of the air, also the deleterious gases mingled with it. The last trouble has been obviated to a considerable extent, but the first still exists.

The advantage claimed for hot water is, it is easily managed, a very important matter, and maintains a very even temperature in the house, etc. The disadvantages are that it requires some time to cool after the radiator is turned off, and is slow in heating up at first, and is liable to freeze up and cause much damage The first cost is also large. Low pressure steam, it is claimed, is quicker in heating up and also quicker in cooling, is more controllable and is fully as safe. First cost is less, and the economy in running fully as great or even greater. Both are adapted to all systems of heating. The combination system of hot air furnace with steam or hot water is often a great aid to the furnace system, for as the furnace sends the heated air from a central point only, it is found very difficult to secure proper distribution. With the aid of steam or hot water this can be done, but with some disadvantage in loss of simplicity, which has always been a strong point in favor of the furnace. The question arises, however, is the gain greater in the aggregate than the loss, or sufficient to warrant the increased expenditure? It is at best a hybrid, though perhaps no more so than any indirect and direct combined steam or hot water apparatus. It is, when all is said, a valuable arrangement in certain cases.

In every hot air apparatus, no matter how arranged, the actuating power for moving the air is heat, which is not very reliable or forceful as applied. The assistance of a power fan, however small, would greatly aid in obviating all difficulties, both for the hot air furnace and the steam and hot water indirect. This accelerator would make the circulation positive and prompt, and by the more rapid conduction of heat and the great reduction in the amount of radiating surface required save a large part of the cost of the power used.

Foreigners often complain of the excessive heat in our houses in the winter. This is to a certain extent true, and there is much that can justly be said about the excessive dryness of the air and want of frequent change by ventilation, but this is being gradually improved. When one has, however, experienced the discomforts, to say the least, of the almost entire absence of heat in the houses in such great cities as Paris, London, Berlin and Vienna in zero weather, and seen the crude means they still use for heating purposes, it makes us feel that in this line of progress we are, as in many other lines, far ahead of the rest of the world.

In England the open fire place is still in vogue, with plenty of smoky coal; this has the advantage of some ventilation, but as the heat radiates from one central point only, the rest of the room to be heated has almost every degree of temperature from the torrid to the frigid zone.

In France, coal being scarce and high, the charcoal brazier is still in some use, and a coal fire is a luxury. In Germany, Russia and Austria the air tight stove is chiefly used. With this apparatus ventilation is ignored, and to secure economy in heat every crevice and crack is stopped up, so there is either an absence of the proper amount of heat for comfort or no ventilation. The best form of heating apparatus for dwelling houses is a difficult question to decide. It must depend largely on circumstances, but in a general way it may be said that a combination of indirect for the first floor, or even the second floor, and direct for the other floors and the off corners hard to reach by hot air pipes, is best. The circulation of the air through the indirect radiators may be aided by an accelerator or fan driven by a small electric, gas, oil, gasoline, water or other motor capable of being regulated by hand or automatically.

CHAPTER XV.

A SUCCESSFUL HEATING JOB.*

Knowing the interest that always attaches to actual work which has been done as compared with work that has only been done on paper, we reprint below a complete specification for a steam heating job in a Brooklyn residence, giving all the particulars of the specification as well as of the agreement, naming the articles specified and giving the prices of all. The only change that we have made is in the names of the owner and contractor, the latter being one of the most prominent heating contracting firms in New York City. The specification was carefully drawn up and the operation of the apparatus, we understand, has been entirely successful. As an example of practical work we believe it will interest many of our readers. The engravings were made from the plans furnished by the architect.

Specification

for a low pressure steam heating apparatus for a residence on Diamond street, near Bedford avenue, Brooklyn

MR. G. M LAWTON, ARCHITECT,
No. 508 Madison ave.,
Brooklyn, N. Y.

Workmanship.—The entire work to be executed in the neatest and best manner and in strict accordance with the rules of the New York Board of Fire Underwriters. Where steam pipes pass through floors or partitions they must be covered with heavy galvanized iron tubes with flanges on each end, floor and ceiling plates, &c.

Material.—All materials to be the best of their several kinds, the fittings of heavy pattern and the pipe the best

* Reprinted from *The Metal Worker*, July 27, 1895, with additions showing the method of calculating the surface. This work was designed and supervised in its execution by Mr. McNeill.

made. Eccentric fittings to be used on steam main in cellar and on all horizontal branches to radiators.

Valves.—All valves to be of Fairbank's make, with composition disk seats. Bodies and trimmings to be of best steam metal, heavy pattern. Radiator valves to be nickel plated, to have neat polished hardwood handle, and to be either angle, opposite angle or gate valves; no globe valve to be used except on main return in cellar. Each radiator on first story to have two (2) valves, for steam and return, 1¼-inch and 1-inch respectively. Radiators above first story to have only one valve each, the same to be 1¼-inch. All cocks to be of best steam metal ground tight.

Cutting, etc.—All cutting, excavating and repairing found necessary in the execution of the work to be done by contractor, and all rubbish promptly removed from the premises.

Testing.—The entire apparatus to be thoroughly tested with 30 pounds per square inch steam pressure, made tight and absolutely noiseless in working under varied pressures. A guarantee to be given against defects of workmanship and material for one year from date of acceptance of work.

System.—Steam at low pressure (not to exceed 10 pounds per square inch) generated by boiler in cellar and by suitable piping supplied to vertical loop direct radiators placed in the various rooms throughout the building. The water of condensation to be noiselessly returned by gravity to the boiler. The system of piping in the cellar will comprise steam and return mains, and the radiators on the first floor will be supplied directly from said mains. The radiators above the first floor will be supplied by risers, each of one pipe and carrying both steam and return water. Run mains in cellar without traps or pockets, with ample inclination in the direction of the flow of their contents, and near to and supported from the beams in cellar ceiling by neat expansion hangers.

Boiler.—Furnish and set in position shown on plans, on proper foundations, a Gorton steam boiler, size No. 1, of latest pattern, together with all parts and appliances necessary or desirable for operating the same. Connect boiler to smoke flue (marked B) by a smoke pipe 7 inches in diameter, made of No. 16 galvanized iron, and with damper rod and lever complete.

Trimmings.—Furnish and attach to boiler the following trimmings, etc.: Low pressure steam gauge with

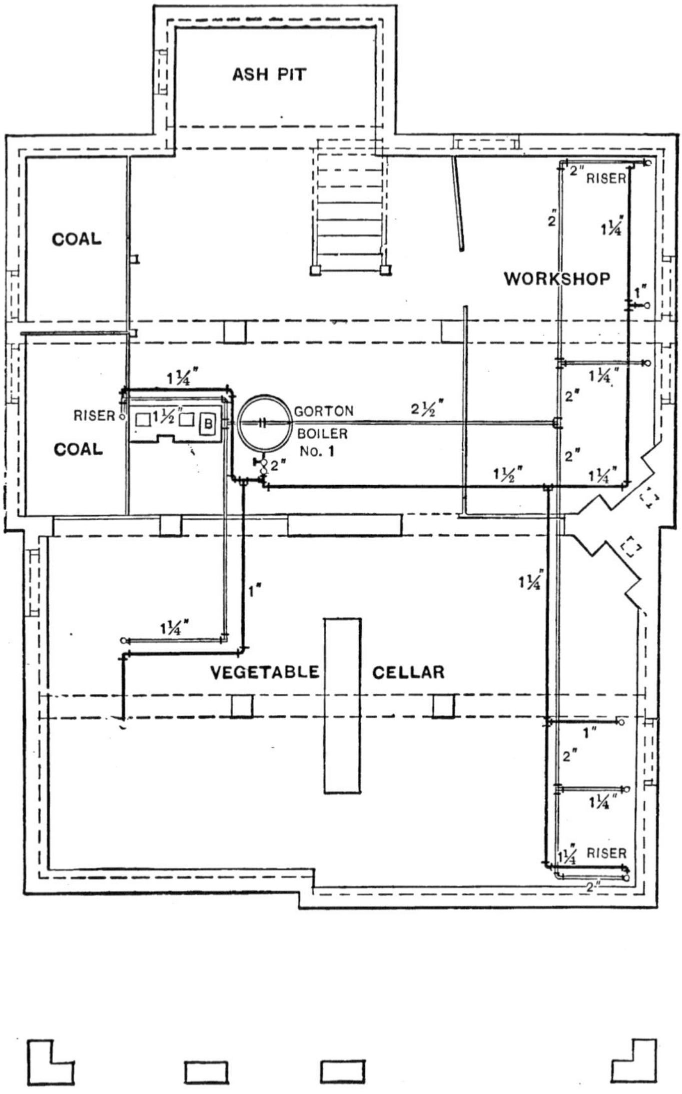
ASH PIT
COAL
WORKSHOP
2" RISER
2"
1¼"
1"
1¼"
2"
2"
1¼"
1½"
1¼"
RISER
1½"
B
GORTON
BOILER
No. 1
2½"
2"
COAL
1¼"
1"
1¼"
VEGETABLE
CELLAR
1"
2"
1¼"
1¼ RISER
2"

Cellar Plan

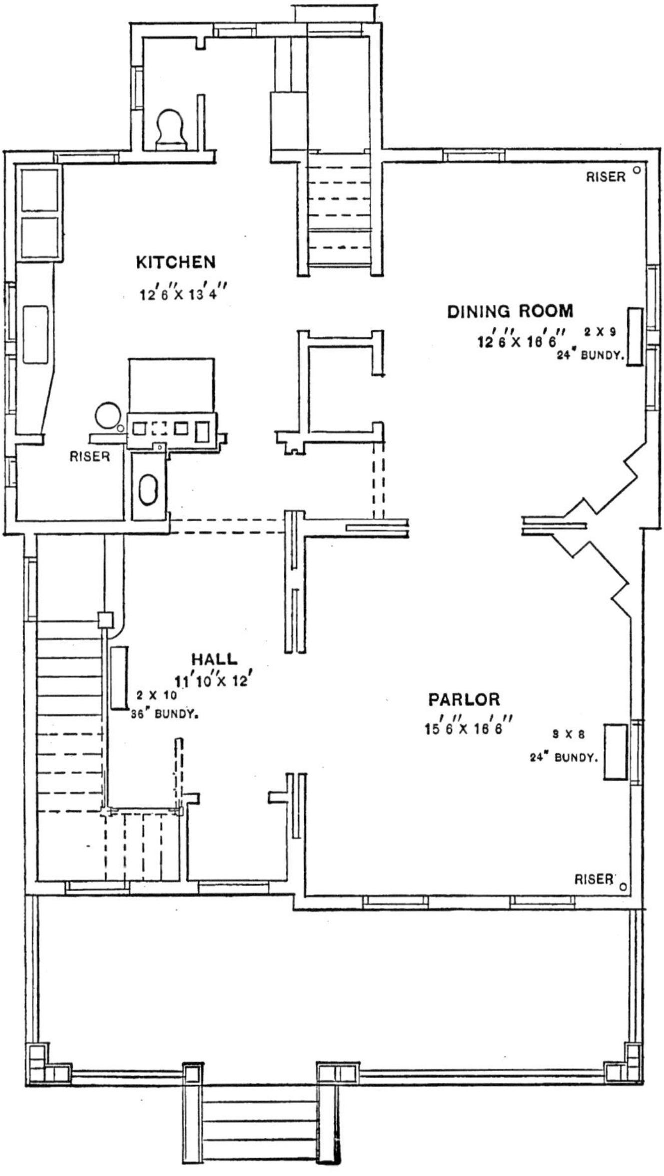

First-Floor Plan.

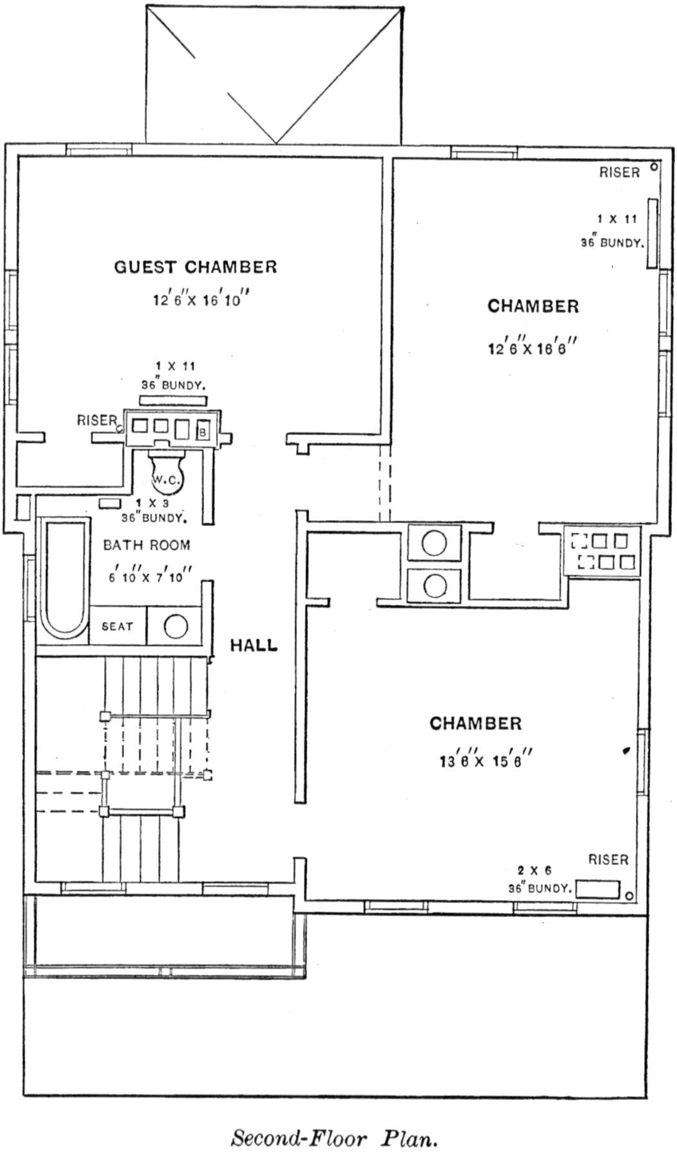

Second-Floor Plan.

siphon; water column with three ½-inch brass gauge cocks and a 12 inch water gauge with brass trimmings; low pressure safety valve, 1¼-inch; regulating bowl with

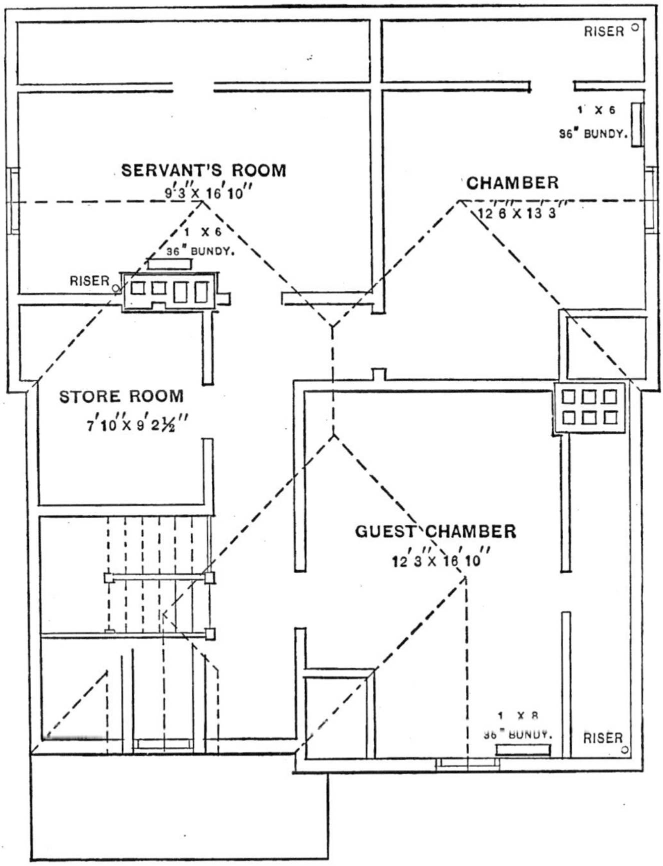

Attic Plan

connection to ash pit door and damper in smoke pipe; blow off cock 1¼-inch connected to sewer; and approved automatic water feeder with hand feed and all necessary cocks; a 2-inch swing check valve and a 2-inch globe valve on main return, and a $\frac{3}{4}$-inch pass-by to sewer with cock. Also furnish a full set of fire tools and a tube brush with jointed rod or cable.

Mains.—Run out from the two openings in the top of boiler for steam outlet two (2) $1\frac{1}{2}$-inch pipes, which unite into one $2\frac{1}{2}$-inch and continue in the direction and of the sizes shown on plans; also make all branch connections to risers on first floor. Run return mains, branches, etc., of the sizes and in the directions as shown on plans. Make connection to bottoms of risers, to radiators on first floors and to boiler below water line in cellar, as shown.

Risers.—Run up from cellar to top story three (3) risers, each of one $1\frac{1}{2}$-inch pipe, with connections to radiators.

Radiators.—Furnish and connect complete in positions shown on plans ten (10) radiators containing in the aggregate 119 Bundy loops. Of these 77 to be 36 inches high and 42 to be 24 inches high. Total, 315 square feet of surface.

Air Valves.—Furnish and attach to each radiator a Van Auken duplex automatic air valve, nickel plated, with drip pipe and a $\frac{3}{8}$-inch pipe run down to cellar.

Bronzing, etc.—All pipes, etc., in cellar to be neatly black japanned. All radiators, risers, etc., above cellar to be neatly gold or silver bronzed.

Articles of Agreement

made this ninth day of July, in the year one thousand eight hundred and ninety-four, between William Brown of the first part, and the Jones & Robinson Company of the second part, in these words:

The said parties of the second part do hereby covenant, promise and agree, to and with the said party of the first part, that they, the said parties of the second part, shall and will, for the consideration hereinafter mentioned, well and sufficiently provide all steam heating materials, and perform all steam heating work, or other materials or work required by the steam heating specification hereto annexed, and also to complete the same on or before the fifteenth day of September, 1894, or as soon as the condition of the building will permit.

And the said party of the first part does hereby covenant, promise and agree, to and with the said parties of the second part, that he, the said party of the first part, shall and will, in consideration of the covenants and agreements being strictly performed and kept by the said parties of the second part, pay unto the said parties of the second part the sum of three hundred and eighty-five dollars ($385) of lawful money, in manner following:

First.—The sum of $100 when all the rising lines are in.

Second.—The sum of $150 when the boiler and all the radiators are delivered.

Third.—The sum of $135, the balance of the contract price, when all the work contracted for is fully completed.

Provided, The said parties of the second part obtain from John Smith, the superintending architect of the works, a written certificate, to the effect that the specification has been complied with, and the work done to his satisfaction and acceptance. And should any dispute arise respecting the construction or meaning of the drawings or specification, the same shall be decided by the said architect.

And it is hereby further agreed by and between the said parties:

First.—The owner may, at any time during the progress of the work, make any desired alteration in the said contract, either by work added thereto, or omitted therefrom, and the same shall in no way affect or make void the contract, but will be added to or deducted from the amount of the contract, as the case may be, by a fair and reasonable valuation.

Second.—Should the contractor, at any time during the progress of the work, refuse or neglect to advance the same consistent with its completion by the time herein specified, or so as to delay or retard other divisions of the work, the owner shall have the power to provide the labor and material necessary to finish the said works, after three days' notice in writing being given, to finish the said works, and the contractor shall be chargeable with the expense, and be liable for any excess over the contract price.

Witness our hands and seals, the day and year first above written.

WM. BROWN. [SEAL.]
THE JONES & ROBINSON CO. [SEAL.]

CALCULATIONS

for a low pressure steam heating apparatus for a residence on Diamond street, near Bedford avenue, Brooklyn, N. Y.

....Owner.

Geo. M. Lawton..Architect

Cellar.—Hight, 7 feet 4 inches, floor to ceiling:

Two-pipe system on the first floor and in cellar, and one-pipe system on the floors above.

First floor.—Hight, 10 feet, floor to ceiling:

	Ratio.	Cubic contents.	Square feet.	Pipe.
Parlor: 15 feet 6 inches x 16 feet	53		48	
6 inches x 10 feet..................	50	2,557	51	24
One Bundy, 3 x 8 = 24 pipes, 24 inches high.				
Dining room: 12 feet x 16 feet	57		36	
6 inches x 10 feet...................	60	2,062	34	18
One Bundy radiator, 2 x 9 = 18 pipes, 24 inches high.				
Main hall: 11 feet 10 inches x 12 feet	23		60	
x 10 feet...........................	25	1,419	57	20
One Bundy, 2 x 10 = 20 pipes, 36 inches high.				

Second floor.—Hight, 9 feet 6 inches, floor to ceiling:

	Ratio.	Cubic contents.	Square feet.	Pipe.
Chamber, front: 13 feet 6 inches x	55		36	
15 feet 6 inches x 9 feet 6 inches...	55	1,987	36	12
One Bundy, 2 x 6 = 12 pipes, 36 inches high.				
Chamber, rear: 12 feet 6 inches x	60		33	
16 feet 6 inches x 9 feet 6 inches...	60	1,959	32	11
One Bundy, 1 x 11 = 11 pipes, 36 inches high.				
Guest chamber: 12 feet 6 inches x	60		33	
16 feet 10 inches x 9 feet 6 inches..	60	1,998	33	11
One Bundy, 1 x 11 = 11 pipes, 36 inches high.				
Bathroom: 6 feet 10 inches x 7 feet	57		9	
10 inches x 9 feet 6 inches.........	55	518	9	3
One Bundy, 1 x 3 = 3 pipes, 36 inches high.				

Attic.—Hight, 9 feet, floor to ceiling:

	Ratio.	Cubic contents.	Square feet.	Pipe.
Guest chamber: 12 feet 3 inches x	76		24	
16 feet 10 inches x 9 feet...........	76	1,846	24	8
1 Bundy, 1 x 8 = 8 pipes, 36 inches high.				
Servant's room: 9 feet 3 inches x	77		18	
16 feet 10 inches x 9 feet...........	80	1,401	17	6
One Bundy, 1 x 6 = 6 pipes, 36 inches high.				
Chamber, rear: 12 feet 6 inches x	83		18	
13 feet 3 inches x 9 feet............	80	1,490	18	6
One Bundy, 1 x 6 = 6 pipes, 36 inches high.				

The double figures shown in the "ratio" and "square feet" columns illustrate the difference which is sometimes necessary in the amount called for and the amount actually put in, caused by the inability to obtain the exact amount of surface in the radiators made and on the market.

[THE END.]

Spotted something that isn't quite right?

Let us know and we'll fix it for future reprints.
And as a thank you for your diligence, we'll
send you a free Sothis book of your choice.

publisher@sothis.press

This page intentionally left blank

Except of course for that line above. And this one.
And whilst we have your attention, perhaps you'd
like to follow us on Instagram @sothis.press. Go on.

Made in United States
North Haven, CT
05 November 2026

21817317R00090